Principles of
Nanotechnology

Principles of
Nanotechnology

MOLECULAR-BASED STUDY OF CONDENSED MATTER IN SMALL SYSTEMS

G. Ali Mansoori
University of Illinois at Chicago, USA

 **World Scientific**

NEW JERSEY • LONDON • SINGAPORE • BEIJING • SHANGHAI • HONG KONG • TAIPEI • CHENNAI

Published by

World Scientific Publishing Co. Pte. Ltd.

5 Toh Tuck Link, Singapore 596224

USA office: 27 Warren Street, Suite 401-402, Hackensack, NJ 07601

UK office: 57 Shelton Street, Covent Garden, London WC2H 9HE

British Library Cataloguing-in-Publication Data
A catalogue record for this book is available from the British Library.

PRINCIPLES OF NANOTECHNOLOGY: MOLECULAR-BASED STUDY OF CONDENSED MATTER IN SMALL SYSTEMS

ISBN 981-256-154-4
ISBN 981-256-205-2 (pbk)

Printed in Singapore.

To Manijeh

Preface

"Principles of Nanotechnology" is a result of my recent research and educational activities concerning the molecular behavior of condensed matter in small systems. It contains a special viewpoint of nano systems based on my prior experiences dealing with the molecular based study of matter, statistical mechanics, thermodynamics and engineering. I have previously presented various components of this subject in workshops, colloquies, seminars and research papers.

I am grateful to numerous colleagues, friends and students, with whom I have sought collaboration and co-authored many research papers. I am specifically thankful to J. Anderson, L. Assoufid, R. Bagherian, B. Chehroudi, D. Dziura, T. Ebtekar, A. Eliassi, K. Esfarjani, T.F. George, A. Johnson, L.A. Kennedy, G.L. Klimchitskaya, C. Megaridess, S. Priyanto, H. Rafii-Tabar, H. Ramezani, D. Samaranski, B. Searles, T.A.F. Soelaiman, G.R. Vakili-Nezhaad, M. Shariaty-Niassar, A. Soltani, B. Soltani, A. Suwono, G. Uslenghi, G. Willing and G. Zhang. These associations have been invaluable in the completion of this book.

The reader should bear in mind that this book serves as a survey of this subject matter. A limited number of topics and references are given and I apologize to those whose publications were omitted.

G.Ali Mansoori
October 8, 2004, Chicago

Contents

Chapter 3 — Thermodynamics and Statistical Mechanics of Small Systems

Chapter 4 — Monte Carlo Simulation Methods for Nanosystems

Chapter 6 — Computer-Based Simulations and Optimizations for Nanosystems

Chapter 7 — Phase Transitions in Nanosystems

Chapter 8 — Positional Assembly of Atoms and Molecules

Chapter 9 — Molecular Self-Assembly

Chapter 1

Advances in Atomic and Molecular Nanotechnology

"Everything we see around us is made of atoms, the tiny elemental building blocks of matter. From stone, to copper, to bronze, iron, steel, and now silicon, the major technological ages of humankind have been defined by what these atoms can do in huge aggregates, trillions upon trillions of atoms at a time, molded, shaped, and refined as macroscopic objects. Even in our vaunted microelectronics of 1999, in our highest-tech silicon computer chip the smallest feature is a mountain compared to the size of a single atom. The resultant technology of our 20th century is fantastic, but it pales when compared to what will be possible when we learn to build things at the ultimate level of control, one atom at a time." **Richard E. Smalley**

Introduction

In this chapter, we present an introduction to the advances made in the atomic and molecular nanotechnology, ability to systematically organize and manipulate properties and behavior of matter in the atomic and molecular levels. It is argued that through nanotechnology, it has become possible to create functional devices, materials and systems on the 1 to 100 nanometer (one billionth of a meter) length scale.

The reasons why nanoscale has become so important are presented. Historical aspects of nanotechnology are introduced starting with the famous 1959 lecture by R.P. Feynman. It is suggested to name the nanometer scale the *Feynman* (ϕnman) *scale* after Feynman's great contributions to nanotechnology (1 *Feynman* $[\phi] \equiv 10^{-9}$ *meter* $= 10^{-3}$ *Micron* $[\mu] = 10$ *Angstrom*s [Å]). Recent inventions and discoveries in atomic and molecular aspects of nanotechnology are presented and the ongoing related research and development activities are introduced.

It is anticipated that the breakthroughs and developments in nanotechnology will be quite frequent in the coming years.

The author of this book has spent over thirty seven years of his adult life researching into the atomic and molecular based study of matter. This has included prediction of the behavior of fluids, solids and phase transitions starting with the consideration of interatomic and intermolecular interactions among atoms and molecules in various phases of matter and phase transitions [1-4]. A few years ago he was introduced to the fascinating subjects of nanoscience and nanotechnology and the fact that it will lead us to the next industrial revolution [5,6]. What you read in this book consist of a reflection of the experiences of the author describing nanoscience and nanotechnology from the atomic and molecular interactions point of view.

If one likes to have the shortest and most complete definition of nanotechnology one should refer to the statement by the US National Science and Technology Council [5] which states: *"The essence of nanotechnology is the ability to work at the molecular level, atom by atom, to create large structures with fundamentally new molecular organization. The aim is to exploit these properties by gaining control of structures and devices at atomic, molecular, and supramolecular levels and to learn to efficiently manufacture and use these devices"*. In short, nanotechnology is the ability to build micro and macro materials and products with atomic precision.

The promise and essence of the nanoscale science and technology is based on the demonstrated fact that materials at the nanoscale have properties (i.e. chemical, electrical, magnetic, mechanical and optical) quite different from the bulk materials. Some of such properties are, somehow, intermediate between properties of the smallest elements (atoms and molecules) from which they can be composed of, and those of the macroscopic materials. Compared to bulk materials, it is demonstrated that nanoparticles possess enhanced performance properties when they are used in similar applications. An important application of nanoparticles is recognized to be the production of a new class of catalysts known as nanocatalysts. Significant advances are being made in this field contributing to the production and detailed understandings of the nature (composition, particle size, and structure) and role of nanoparticles as catalysts in enhancement of chemical reactions. This is because a catalyst performance is a strong function of its particles sizes and size distribution. Surface morphology, surface to volume ratio, and electronic properties of materials could change appreciably due to particle size changes. For instance, it is observed that the heat of adsorption of CO on Ni catalyst and the activation energy for CO dissociation, both, change with decreasing the size of Ni particles in the well-known Fischer-Tropsch synthesis of light hydrocarbons from synthesis gas (a mixture of CO and H_2) [7]. There are many present and expected applications of nanoscience and nanotechnology including bottom-up technology (such as self-replication and self-assembly), microbiological, energy conversion, medical, pharmaceutical, etc, which are rapidly increasing.

The Importance of Nanoscale

The Greek word "nano" (meaning dwarf) refers to a reduction of size, or time, by 10^{-9}, which is one thousand times smaller than a micron. One nanometer (*nm*) is one billionth of a meter and it is also equivalent to ten Angstroms. As such a nanometer is 10^{-9} meter and it is 10,000 times smaller than the diameter of a human hair. A human hair diameter is about 50 micron (i.e., 50×10^{-6} meter) in size, meaning that a 50 nanometer object is about $1/1000^{th}$ of the thickness of a hair. One cubic nanometer (*nm^3*) is roughly 20 times the volume of an individual atom. A nanoelement compares to a basketball, like a basketball to the size of the earth. Figure 1 shows various size ranges for different nanoscale objects starting with such small entities like ions, atoms and molecules.

Size ranges of a few nanotechnology related objects (like nanotube, single-electron transistor and quantum dot diameters) are also shown in this figure. It is obvious that nanoscience, nanoengineering and nanotechnology, all deal with very small sized objects and systems.

Officially, the United States National Science Foundation [6] defines nanoscience / nanotechnology as studies that deal with materials and systems having the following key properties.

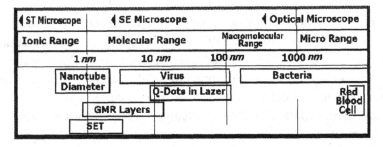

◀ ST Microscope	◀ SE Microscope		◀ Optical Microscope
Ionic Range	Molecular Range	Macromolecular Range	Micro Range
1 *nm*	10 *nm*	100 *nm*	1000 *nm*

Figure 1. Comparison of size ranges for several entities as compared to some nanotechnology devices: SET (Single-electron transistor), GMR (Giant magneto resistive), Q-DOTS (Quantum dots). SE stands for Scanning Electron and ST stands for Scanning Tunneling.

(1) Dimension: at least one dimension from 1 to 100 nanometers (*nm*).
(2) Process: designed with methodologies that shows fundamental control over the physical and chemical attributes of molecular-scale structures.
(3) Building block property: they can be combined to form larger structures. Nanoscience, in a general sense, is quite natural in microbiological sciences considering that the sizes of many bioparticles dealt with (like enzymes, viruses, etc) fall within the nanometer range.

Nanoscale is a magical point on the dimensional scale: Structures in nanoscale (called *nanostructures*) are considered at the borderline of the smallest of human-made devices and the largest molecules of living systems. Our ability to control and manipulate nanostructures will make it possible to exploit new physical, biological and chemical properties of systems that are intermediate in size, between single atoms, molecules and bulk materials.

There are many specific reasons why nanoscale has become so important some of which are as the following [6]:

(i) The quantum mechanical (wavelike) properties of electrons inside matter are influenced by variations on the nanoscale. By nanoscale design of materials it is possible to vary their micro and macroscopic properties, such as charge capacity, magnetization and melting temperature, without changing their chemical composition.

(ii) A key feature of biological entities is the systematic organization of matter on the nanoscale. Developments in nanoscience and nanotechnology would allow us to place man-made nanoscale things inside living cells. It would also make it possible to make new materials using the self-assembly features of nature. This certainly will be a powerful combination of biology with materials science.

(iii) Nanoscale components have very high surface to volume ratio, making them ideal for use in composite materials, reacting systems, drug delivery, and chemical energy storage (such as hydrogen and natural gas).

(iv) Macroscopic systems made up of nanostructures can have much higher density than those made up of microstructures. They can also be better conductors of electricity. This can result in new electronic device concepts, smaller and faster circuits, more sophisticated functions, and greatly reduced power consumption simultaneously by controlling nanostructure interactions and complexity.

Atomic and Molecular Basis of Nanotechnology

The molecular theory of matter starts with quantum mechanics and statistical mechanics. According to the quantum mechanical Heisenberg Uncertainty Principle the position and momentum of an object cannot simultaneously and precisely be determined [8]. Then the first question that may come into mind is, how could one be able to brush aside the Heisenberg Uncertainty Principle, Figure 2, to work at the atomic and molecular level, atom by atom as is the basis of nanotechnology.

The Heisenberg Uncertainty Principle helps determine the size of electron clouds, and hence the size of atoms. According to Werner Heisenberg "The more precisely the POSITION is determined, the less precisely the MOMENTUM is known". Heisenberg's Uncertainty

Principle applies only to the subatomic particles like electron, positron, photon, etc. It does not forbid the possibility of nanotechnology, which has to do with the position and momentum of such large particles like atoms and molecules. This is because the mass of the atoms and molecules are quite large and the quantum mechanical calculation by the Heisenberg Uncertainty Principle places no limit on how well atoms and molecules can be held in place [8].

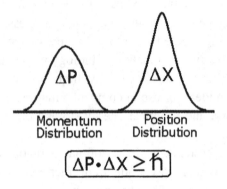

Figure 2. Heisenberg Uncertainty Principle.

Although we have long been aware of, and many investigators have been dealing with, "nano" sized entities; the historic birth of the nanotechnology is commonly credited to Feynman. Historically nanotechnology was for the first time formally recognized as a viable field of research with the landmark lecture delivered by Richard P. Feynman, the famous Noble Laureate physicist on December 29th 1959 at the annual meeting of the American Physical Society [9]. His lecture was entitled *"There's Plenty of Room at the Bottom* - An invitation to enter a new field of physics". Feynman stated in his lecture that the entire encyclopedia of Britannica could be put on the tip of a needle and, in principle, there is no law preventing such an undertaking. Feynman described then the advances made in this field in the past and he envisioned the future for nanotechnology. His lecture was published in the February 1960 issue of Engineering & Science quarterly magazine of California Institute of Technology.

In his talk Feynman also described how the laws of nature do not limit our ability to work at the molecular level, atom by atom. Instead, he said, it was our lack of the appropriate equipment and techniques for doing so. Feynman in his lecture talked about "*How do we write small?*", "*Information on a small scale*", possibility to have "*Better electron microscopes*" that could take the image of an atom, doing things small scale through "*The marvelous biological system*", "*Miniaturizing the computer*", "*Miniaturization by evaporation*" example of which is thin film formation by chemical vapor deposition, solving the "*Problems of lubrication*" through miniaturization of machinery and nanorobotics, "*Rearranging the atoms*" to build various nanostructures and nanodevices, and behavior of "*Atoms in a small world*" which included atomic scale fabrication as a bottom-up approach as opposed to the top-down approach that we are accustomed to [10]. Bottom-up approach is self-assembly of machines from basic chemical building blocks which is considered to be an ideal through which nanotechnology will ultimately be implemented. Top-down approach is assembly by manipulating components with much larger devices, which is more readily achievable using the current technology.

It is important to mention that almost all of the ideas presented in Feynman's lecture and even more, are now under intensive research by numerous nanotechnology investigators all around the world. For example, in his lecture Feynman challenged the scientific community and set a monetary reward to demonstrate experiments in support of miniaturizations. Feynman proposed radical ideas about miniaturizing printed matter, circuits, and machines. "*There's no question that there is enough room on the head of a pin to put all of the Encyclopedia Britanica*" he said. He emphasized "*I'm not inventing antigravity, which is possible someday only if the laws (of nature) are not what we think*" He added "*I am telling what could be done if the laws are what we think; we are not doing it simply because we haven't yet gotten around to it.*" Feynman's challenge for miniaturization and his unerringly accurate forecast was met forty years later, in 1999, [11] by a team of scientists using a nanotechnology tool called Atomic Force Microscope (AFM) to perform Dip Pen Nanolithography (DPN) the result of which is shown in Figure 3. In this DPN an AFM tip is simply coated with molecular 'ink' and then brought in contact with the surface to be patterned. Water condensing from the immediate environment forms a capillary between the AFM tip and the surface. Work is currently under way to investigate the potential of the DPN technique as more than a quirky tool for nanowriting, focusing on applications in microelectronics, pharmaceutical screening, and biomolecular sensor technology.

Feynman in 1983 talked about a scaleable manufacturing system, which could be made to manufacture a smaller scale replica of itself [12]. That, in turn would replicate itself in smaller scale, and so on down to molecular scale. Feynman was subscribing to the "Theory of Self-Reproducing Automata" proposed by von Neumann the 1940's eminent mathematician and physicist who was interested in the question of whether a machine can self-replicate, that is, produce copies of itself (see [13] for details). The study of man-made self-replicating systems has been taking place now for more than half a century. Much of this work is motivated by the desire to understand the fundamentals involved in self-replication and advance our knowledge of single-cell biological self-replications.

Some of the other recent important achievements about which Feynman mentioned in his 1959 lecture include the manipulation of single atoms on a silicon surface [14], positioning single atoms with a scanning tunneling microscope [15] and the trapping of single, 3 *nm* in diameter, colloidal particles from solution using electrostatic methods [16].

In early 60's there were other ongoing research on small systems but with a different emphasis. A good example is the publication of two books on "Thermodynamics of Small Systems" by T.L. Hill [17] in early 1960s. Thermodynamics of small systems is now called "nanothermodynamics" [18]. The author of this book was privileged to offer a short course on nanothermodynamics to a large group of university professors, other scientists and graduate students during last May [19].

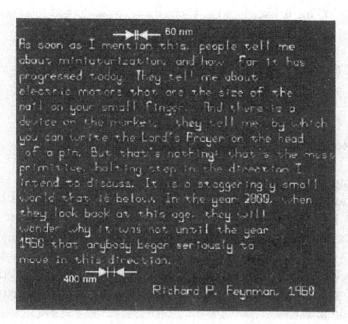

Figure 3. Demonstration of lithography miniaturization challenge by the scientists at the Northwestern University [11] as predicted by Feynman in 1959 using an AFM tip to write a paragraph of nanometer-sized letters with a single layer of mercaptohexadecanoic acid on a gold surface. Contrast is enhanced by surrounding each letter with a layer of a second "ink"--octadecanethiol.

In 1960s when Feynman recognized and recommended the importance of nanotechnology the devices necessary for nanotechnology were not invented yet. At that time, the world was intrigued with space exploration, discoveries and the desire and pledges for travel to the moon, partly due to political rivalries of the time and partly due to its bigger promise of new frontiers that man had also not captured yet. Research and developments in small (nano) systems did not sell very well at that time with the governmental research funding agencies and as a result the scientific community paid little attention to it.

It is only appropriate to name the nanometer scale "the *Feynman* (ϕnman) *scale*" after Feynman's great contribution and we suggest the notation "ϕ" for it like $\mathring{A}$ as used for Angstrom scale and μ as used for micron scale.

One *Feynman* (ϕ) $\equiv 1$ *Nanometer* (*nm*)= 10 *Angstroms* ($\mathring{A}$)= 10^{-3}
Micron (μ) = 10^{-9} *Meter* (*m*)

Some Recent Key Inventions and Discoveries

Scanning Tunneling Microscope: Nanotechnology received its greatest momentum with the invention of scanning tunneling microscope (STM) in 1985 by Gerd K. Binnig and Heinrich Rohrer, staff scientists at the IBM's Zürich Research Laboratory [20]. That happened forty-one years after Feynman's predictions. To make headway into a realm of molecule-sized devices, it would be necessary to survey the landscape at that tiny scale. Binning and Rohrer's scanning tunneling microscope offered a new way to do just that.

STM allows imaging solid surfaces with atomic scale resolution. It operates based on tunneling current, which starts to flow when a sharp tip is mounted on a piezoelectric scanner approaches a conducting surface at a distance of about one *nm* (1 ϕ). This scanning is recorded and displayed as an image of the surface topography. Actually the individual atoms of a surface can be resolved and displayed using STM.

Atomic Force Microscope: After the Nobel Prize award in 1986 to Binnig and Rohrer for the discovery of STM it was quickly followed by the development of a family of related techniques which, together with STM, may be classified in the general category of Scanning Probe Microscopy (SPM) techniques. Of the latter technologies, the most important is undoubtedly the atomic force microscope (AFM) developed in 1986 by Binnig, Quate and Gerber [21]. Figure 4 shows schematic of two typical AFMs.

An AFM, as shown in Figure 4, is a combination of the principle of STM and the stylus profilometer. It enables us to study non-conducting surfaces, because it scans van der Waals forces with its "atomic" tips. Presently several vendors are in the market with commercial AFMs.

AFM and STM possess three-dimensional resolutions up to the atomic scale which cannot be met by any other microscope. The AFMs sold by most manufacturers are generally user-friendly and they produce detailed images. AFM has found versatile applications in nanotechnology as well as other fields of science and engineering. The main components of this tool are thin cantilever with extremely sharp (1-10 nm [ϕ] in radius) probing tip, a 3D piezo-electric scanner, and optical system to measure deflection of the cantilever. When the tip is brought into contact with the surface or in its proximity, or is tapping the surface, it being affected by a combination of the surface forces (attractive and repulsive). Those forces cause cantilever bending and torsion, which is continuously, measures via the deflection of the reflected laser beam.

3D scanner moves the sample or, in alternative designs, the cantilever, in 3 dimensions thus scanning predetermined area of the

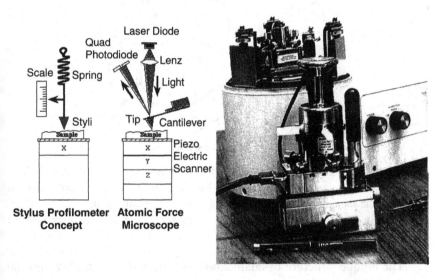

Figure 4. Schematic of a typical AFM and its function as compared with a stylus profilometer. As it is shown an AFM has similarities to a conventional stylus profilometer, but with a much higher resolution in nano scale. In the right hand side pictures of two AFMs are shown.

surface. A vertical resolution of this tool is extremely high, reaching 0.01 *nm* [ϕ], which is on the order of atomic radius.

Diamondoids: The smallest diamondoid molecule was first discovered and isolated from a Czechoslovakian petroleum in 1933. The isolated substance was named adamantane, from the Greek for diamond. This name was chosen because it has the same structure as the diamond lattice, highly symmetrical and strain free as shown in Figure 5. It is generally accompanied by small amounts of alkylated adamantanes: 2-methyl-; 1-ethyl-; and probably 1-methyl-; 1,3-dimethyl; and others. From the bionanotechnology point of view diamondoids are in the category of organic nanostructures.

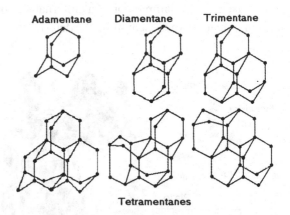

Adamentane   Diamentane   Trimentane

Tetramentanes

Figure 5. Chemical structures of diamondoid molecules. These compounds have diamond-like fused ring structures which can have many applications in nanotechnology. They have the same structure as the diamond lattice, i.e., highly symmetrical and strain free. The rigidity, strength and assortment of their 3-d shapes make them valuable molecular building blocks.

The unique structure of adamantane is reflected in its highly unusual physical and chemical properties. The carbon skeleton of adamantane comprises a small cage structure. Because of this, adamantane and diamondoids in general are commonly known as **cage hydrocarbons**. In a broader sense they may be described as saturated, polycyclic, cage-like hydrocarbons. The diamond-like term arises from the fact that their carbon atom structure can be superimposed upon a diamond lattice. The simplest of these polycyclic diamondoids is adamantane, followed by its homologues diamantane, tria-, tetra-, penta- and hexamantane.

Diamondoids have diamond-like fused ring structures which can have applications in nanotechnology. They have the same structure as the diamond lattice, i.e., highly symmetrical and strain free. Diamondoids offer the possibility of producing variety of nanostructural shapes. We expect them to have the potential to produce possibilities for application as molding and cavity formation characteristics due to their organic nature and their sublimation potential. They have quite high strength, toughness, and stiffness compared to other known molecule.

Diamondoids are recently named as the building blocks for nanotechnology [22]. Here is a partial list of applications of diamondoids in Nanotechnology and other fields [23]:

- Antiviral drug

- Cages for drug delivery

- Designing molecular capsules

- Drug Targeting

- Gene Delivery

- In designing an artificial red blood cell, called "Respirocyte"

- In host-guest chemistry and combinatorial chemistry

- In Nanorobots

- Molecular machines

- Molecular Probe

- Nanodevices

- Nanofabrication

• Nanomodule

• Organic molecular building blocks in formation of nanostructures

• Pharmacophore-based drug design

• Positional assembly

• Preparation of fluorescent molecular probes

• Rational design of multifunctional drug systems and drug carriers

• Self-assembly: DNA directed self-assembly

• Shape-targeted nanostructures

• Synthesis of supramolecules with manipulated architecture

• Semiconductors which show a negative electron affinity

Buckyballs: By far the most popular discovery in nanotechnology is the Buckminsterfullerene molecules. Buckminsterfullerene (or fullerene), C_{60}, as is shown in Figure 6 is another allotrope of carbon (after graphite and diamond), which was discovered in 1985 by Kroto and collaborators [24]. These investigators used laser evaporation of graphite and they found C_n clusters (with n>20 and even-numbers) of which the most common were found to be C60 and C70. For this discovery by Curl, Kroto and Smalley were awarded the 1996 Nobel Prize in Chemistry. Later fullerenes with larger number of carbon atoms (C_{76}, C_{80}, C_{240}, etc.) were also discovered.

Since the time of discovery of fullerenes over a decade and a half ago, a great deal of investigation has gone into these interesting and unique nanostructures. They have found tremendous applications in nanotechnology. In 1990 a more efficient and less expensive method to produce fullerenes was developed by Krätchmer and collaborators [25]. Further research on this subject to produce less expensive fullerene is in progress [26]. Availability of low cost fullerene will pave the way for

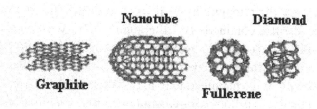

Figure 6. The four allotropes of carbon.

further research into practical applications of fullerene and its role in nanotechnology.

Carbon Nanotubes: Carbon nanotubes were discovered by Iijima in 1991 [27] using an electron microscope while studying cathodic material deposition through vaporizing carbon graphite in an electric arc-evaporation reactor under an inert atmosphere during the synthesis of Fullerenes [28]. The nanotubes produced by Iijima appeared to be made up of a perfect network of hexagonal graphite, Figure 6, rolled up to form a hollow tube.

The nanotube diameter range is from one to several nanometers which is much smaller than its length range which is from one to a few micrometers. A variety of manufacturing techniques has since been developed to synthesize and purify carbon nanotubes with tailored characteristics and functionalities. Controlled production of single-walled carbon nanotubes is one of the favorite forms of carbon nanotube which has many present and future applications in nanoscience and nanotechnology. Laser ablation chemical vapor deposition joined with metal-catalyzed disproportionation of suitable carbonaceous feedstock are often used to produce carbon nanotubes [29-31]. Figure 7 is the scanning electron microscope (SEM) images of a cluster of nanotubes recently produced through plasma enhanced chemical vapor deposition at two different temperatures [29].

Carbon nanotubes and fullerenes are shown to exhibit unusual photochemical, electronic, thermal and mechanical properties [32-35]. It is also shown that single-walled carbon nanotubes (SWCNTs) could behave metallic, semi-metallic, or semi-conductive one-dimensional objects [32], and their longitudinal thermal conductivity could exceed the in-plane thermal conductivity of graphite [33]. Very high tensile strength (~100 times that of steel) of ropes made of SWCNTs has recently been determined experimentally [34]. When dispersed in another medium, it is demonstrated that SWCNTs could retain their intrinsic mechanical attributes or even augment the structural properties of their medium host [35]. SWCNTs have similar electrical conductivity as copper and similar thermal conductivity as diamond.

There is a great deal of interest and activity in the present day to find applications for fullerene and carbon nanotube. There are many ongoing research activities to understand the characteristics of carbon nanotubes including their physicochemical properties, their stability and

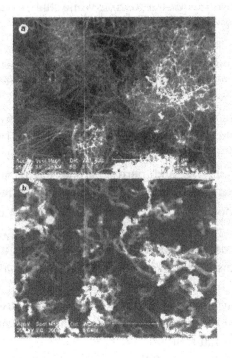

Figure 7. Carbon nanotubes produced using plasma-enhanced chemical vapor deposition at various temperatures [29]. SEM images of deposited carbon nanotubes at (a) 650 °C, (b) 700 °C.

behavior under stress and strain, their interactions with other molecules and nanostructures and their utility for novel applications [36-40].

Cyclodextrins, Liposome and Monoclonal Antibody: At the same time that chemists, materials scientists and physicists have been experimenting with structures like carbon nanotubes and buckyballs and diamondoids, biologists have been making their own advances with other nanoscale structures like cyclodextrins [41], liposomes [42] and monoclonal antibodies [43]. These biological nanostructures have many applications including drug delivery and drug targeting.

Cyclodextrins, as shown in Figure 8, are cyclic oligosaccharides. Their shape is like a truncated cone and they have a relatively hydrophobic interior. They have the ability to form inclusion complexes with a wide range of substrates in aqueous solutions. This property has led to their application for encapsulation of drugs in drug delivery.

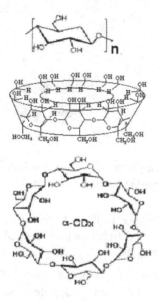

Figure 8. Chemical formula and structure of Cyclodextrins – For n=6 it is called α-CDx, n=7 is called β-CDx, n=8 is called γ-CDx. Cyclodextrins are cyclic oligosaccharides. Their shape is like a truncated cone and they have a relatively hydrophobic interiors. They have the ability to form inclusion complexes with a wide range of substrates in aqueous solution. This property has led to their application for encapsulation of drugs in drug delivery.

Liposome is a spherical synthetic lipid bilayer vesicle, created in the laboratory by dispersion of a phospholipid in aqueous salt solutions. Liposome is quite similar to a micelle with an internal aqueous compartment. Liposomes, which are in nanoscale size range, as shown in Figure 9, self-assemble based on hydrophilic and hydrophobic properties and they encapsulate materials inside. Liposome vesicles can be used as carriers for a great variety of particles, such as small drug molecules, proteins, nucleotides and even plasmids to tissues and into cells. For example, a recent commercially available anticancer drug is a liposome, loaded with doxorubicin, and is approximately 100-nanometer in diameter.

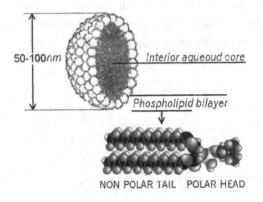

Figure 9. Cross section of a liposome - a synthetic lipid bilayer vesicle that fuses with the outer cell membrane and is used to transport small molecules to tissues and into cells.

A monoclonal antibody protein molecule consists of four protein chains, two heavys and two lights, which are folded to form a Y-shaped structure (see Figure 10). It is about ten nanometers in diameter. This small size is important, for example, to ensure that intravenously administered these particles can penetrate small capillaries and reach cells in tissues where they are needed for treatment. Nanostructures smaller than 20 *nm* can transit out of blood vessels.

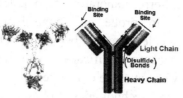

Antibody & its Structure

Figure 10. An antibody is a protein (also called an immunoglobulin) that is manufactured by lymphocytes (a type of white blood cell) to neutralize an antigen or foreign protein.

Ongoing Research and Development Activities

The atomic-scale and cutting-edge field of nanotechnology which is considered to lead us to the next industrial revolution is likely to have a revolutionary impact on the way things will be done, designed and manufactured in the future.

Nanotechnology is entering into all aspects of science and technology including, but not limited to, aerospace, agriculture, bioengineering, biology, energy, the environment, materials, manufacturing, medicine, military science and technology. It is truly an atomic and molecular manufacturing approach for building chemically and physically stable structures one atom or one molecule at a time. Presently some of the active nanotechnology research areas include nanolithography, nanodevices, nanorobotics, nanocomputers, nanopowders, nanostructured catalysts and nanoporous materials, molecular manufacturing, diamondoids, carbon nanotube and fullerene products, nanolayers, molecular nanotechnology, nanomedicine, nanobiology, organic nanostructures to name a few.

We have known for many years that several existing technologies depend crucially on processes that take place on the nanoscale. Adsorption, lithography, ion-exchange, catalysis, drug design, plastics and composites are some examples of such technologies. The "nano" aspect of these technologies was not known and, for the most part, they were initiated accidentally by mere luck. They were further developed using tedious trial-and-error laboratory techniques due to the limited ability of the times to probe and control matter on nanoscale. Investigations at nanoscale were left behind as compared to micro and macro length scales because significant developments of the nanoscale investigative tools have been made only recently.

The above mentioned technologies, and more, stand to be improved vastly as the methods of nanotechnology develop. Such methods include the possibility to control the arrangement of atoms inside a particular molecule and, as a result, the ability to organize and control matter simultaneously on several length scales. The developing concepts of nanotechnology seem pervasive and broad. It is expected to influence every area of science and technology, in ways that are clearly unpredictable.

Nanotechnology will also help solve other technology and science problems. For example, we are just now starting to realize the benefits that nanostructuring can bring to,

(a) wear-resistant tires made by combining nanoscale particles of inorganic clays with polymers as well as other nanoparticle reinforced materials,

(b) greatly improved printing brought about by nanoscale particles that have the best properties of both dyes and pigments as well as advanced ink jet systems,

(c) vastly improved new generation of lasers, magnetic disk heads, nanolayers with selective optical barriers and systems on a chip made by controlling layer thickness to better than a nanometer,

(d) design of advanced chemical and bio-detectors,

(e) nanoparticles to be used in medicine with vastly advanced drug delivery and drug targeting capabilities,

(f) chemical-mechanical polishing with nanoparticle slurries, hard coatings and high hardness cutting tools.

The following selected observations regarding the expected future advances are also worth mentioning at this juncture [6]:

(A) The most complex arrangements of matter known to us are those of living entities and organs. Functions of living organisms depend on specific patterns of matter on all various length scales. Methods of nanotechnology could provide a new dimension to the control and improvement of living organisms.

(B) Photolithographic patterning of matter on the micro scale has led to the revolution in microelectronics over the past few decades. With nanotechnology, it will become possible to control matter on every important length scale, enabling tremendous new power in materials design.

(C) Biotechnology is expected to be influenced by nanotechnology greatly in a couple of decades. It is anticipated that, for example, this will revolutionize healthcare to produce ingestible systems that will be harmlessly flushed from the body if the patient is healthy but will notify a physician of the type and location of diseased cells and organs if there are problems.

(D) Micro and macro systems constructed of nanoscale components are expected to have entirely new properties that have never before been identified in nature. As a result, by altering and design of the structure of materials in the nanoscale range we would be able to systematically and appreciably modify or change selected properties of matter at macro and micro scales. This would include, for example, production of polymers or composites with most desirable properties which nature and existing technologies are incapable of producing.

(E) Robotic spacecraft that weigh only a few pounds will be sent out to explore the solar system, and perhaps even the nearest stars. Nanoscale traps will be constructed that will be able to remove pollutants from the environment and deactivate chemical warfare agents. Computers with the capabilities of current workstations will be the size of a grain of sand and will be able to operate for decades with the equivalent of a single wristwatch battery.

(F) There are many more observations in the areas of inks and dyes, protective coatings, dispersions with optoelectronic properties, nanostructured catalysts, high reactivity reagents, medicine, electronics, structural materials, diamondoids, carbon nanotube and fullerene products and energy conversion, conservation, storage and usage which are also worth mentioning.

(G) Many large organic molecules are known to forming organic nanostructures of various shapes as shown in Figures 5 and 11 the

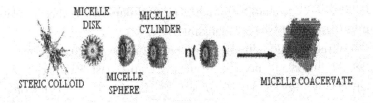

Figure 11. Organic nanostructure self-assemblies of various shapes [44,45].

deriving force of which is the intermolecular interaction energies between such macromolecules [44-46].

There has been an appreciable progress in research during the past few years on organic nanostructures, such as thin film nanostructures, which have excellent potential for use in areas that are not accessible to more conventional, inorganic nanostructures. The primary attraction of organic nanostructures is their potential for molding, coating, and the extreme flexibility that they have in being tailored to meet the needs of a particular application. The organic nanostructure materials are easily integrated with conventional inorganic nanostructures (like semiconductor devices), thereby providing additional functionality to existing photonic circuits and components. Some progress has been made in understanding the formation and behavior of organic nanostructures that might be formed to serve as elements of nanomaterials and also on synthetic strategies for creating such structures [44-46]. The ultimate goal is to achieve a better understanding of the fundamental molecular processes and properties of these nanostructures which are dominated by grain boundaries and interfaces. In understanding the behavior and the properties of these nanostructures the potential for technological applications will be considered.

Figure 12 demonstrates a number of the major expected future and few present activities and possibilities resulting from advances in nanotechnology. According to this figure the impact of implementation of nanotechnology is quite broad. The list of possibilities is expanding quite rapidly.

The impacts of nanotechnology advances are being felt in broad areas of science and technology. It should be pointed out that

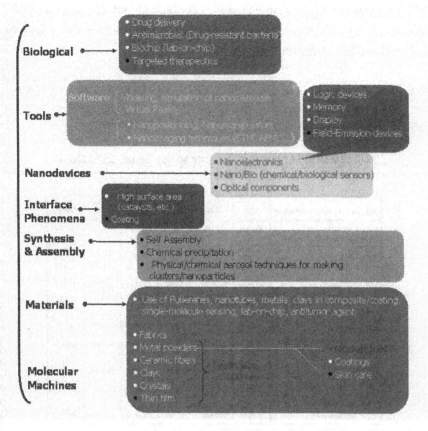

Figure 12. Some of the expected future products and possibilities resulting from the advances in nanotechnology.

nanoelectronic, nanolithography, nanosensors and drug delivery industries have a much clear and distinguished future.

Table 1 demonstrates the expected trend in nanolithography advances indicating various technologies under investigation for development of pilot and production lines of ICs (integrated circuits) [47]. As an example, the half pitch of the dynamic random access memory (DRAM), a type of memory used in most personal computers, (i.e., smallest feature size) is expected to go below the 100 *nm* [ϕ] mark by about 2005. Unpredictable adverse economical factors could probably delay the pace of such developments and hence affect the projected milestone dates shown in Table 1.

Table 1. The International Technology Roadmap for Semiconductors' goals, or "nodes," for rapid decreases in chip size and increases in computer-processor speed (from http://www.lbl.gov/Science-Articles/Archive/ALS-EUVL-sidebar.html).

International Technology Roadmap for Semiconductors*

First year of volume production	2001	2003* 2004	2005* 2007	2007* 2010	2009* 2013	2011* 2016
Technology Generation (Dense lines, printed in resist)	130 nm	90 nm	65 nm	45 nm	32 nm	22 nm
Isolated Lines (in resist) [Physical gate, post-etch]	90 nm [65 nm]	53 nm [37 nm]	35 nm [25 nm]	25 nm [18 nm]	18 nm [13 nm]	13 nm [9 nm]
Chip Frequency	1.7 GHz	4.0 GHz	6.8 GHz	12 GHz	19 GHz	29 GHz
Transistors per chip (HV) (3 x for HP ; 5 x for ASICs)	100 M	190 M	380 M	780 M	1.5 B	3.1 B
DRAM Memory (bits)	510 M	1.1 G	4.3 G	8.6 G	34 G	69 G
Gate CD Control (3σ, post-etch)	5 nm	3 nm	2 nm	1.5 nm	1.1 nm	0.7 nm
Field Size (mm x mm)	25×32	25×32	22×26	22×26	22×26	22×26
Chip Size (mm) (2.2 x for HP ; to 4 x for ASIC)	140	140	140	140	140	140
Wafer Size (diameter)	300 mm	300 mm	300 mm	450 mm	450 mm	450 mm

*Semiconductor Industry Association (SIA), December 2001. *Possible 2-year cycle.

As of the year 2000, the pursuit of mass-produced and inexpensive miniaturization process by the semiconductor industry was reduced the line widths on IC chips down to about 100 ϕ [nm] via extension of the standard photolithography [48]. It is argued that since the wavelength of laser light source determines the width of the smallest line that could be formed on wafer, to pattern ever-finer lines by the use of photolithography, the industry is now making transition from krypton-fluoride excimer laser light source, with wavelength of 248 ϕ [nm], to argon-fluoride lasers emitting at 193 ϕ [nm] wavelength. Within a few years, even argon-fluoride wavelength will be too long and the Fluoride lasers at 157 ϕ [nm] wavelength would be needed. This should bring about serious optical-property-of-materials challenges, as most common optical materials do not transmit at this low wavelength. Below the fluoride laser wavelength lays the transition to extreme ultraviolet range, known as "soft" X-rays, with a wavelength of about 13 ϕ [nm]. The Semiconductor Industry Association periodically outlines future prospects and challenges in a technology "roadmap." Anticipated milestones are called "nodes," defined as a bi-yearly reduction in the size of commercially manufactured chips leading to an increase in chip density by a factor of two. Nodes are expressed as distances in nanometers, half the distance, or pitch, between lines in a set of equally wide, equally spaced lines etched into a chip like a DRAM or MPU (microprocessor unit). Nodes at 90, 65, 45, and 32 ϕ [nm] are expected to be reached within the decade. In nanotechnology this method of fabrication (i.e., photolithography) is referred to as "top-down" approach. The standard microelectronic fabrication for inexpensive and mass-produced integrated circuit chips could reach to its limit and then we speak of nanofabrication, something not yet converged and subject of intense research.

Many other unpredictable advances resulting from nanotechnology are inevitable. Thus, the future prospects for nanotechnology actually represent a revolutionary super-cutting-edge field that is expected to eventually become the foundation for many disparate areas that we cannot even foresee at this time. It is then no wonder that it is considered to lead the humanity to the next industrial revolution.

Future Prospects in Nanoscience and Nanotechnology

Presently nanotechnology and its associated research discipline of nanoscience, together, constitute the complete spectrum of activities towards the promised next industrial revolution. They span the whole spectrum of physical, chemical, biological and mathematical sciences needed to develop the purposeful capabilities of nano manipulations, nano structural modifications, miniaturization and bottom-up technology originally proposed by Richard Feynman in his well-known 1959 lecture. The emerging fields of nanoscience and nanotechnology are also creating the necessary experimental and computational tools for the design and fabrication of nano-dimension electronic, photonic, biological and energy transfer components, such as quantum dots, atomic wires, operating on nanoscopic length scales, etc.

Nanoscience and nanotechnology should have major impacts on several key scientific and technological activities in a not too distant future. Expansions on these subjects will have a lot to do on the technological advances in instruments and tools of fabrication and manipulation in nano scale. Such instruments and tools are the means for live visualization and manipulation in nano world. They are presently expensive and, as a result, not available to many investigators. Technological advances are always followed with reduction of prices as has been the case with the electronic and communication industry products in recent decades.

The decisive and important leading role of molecular-based techniques for the study of matter in the fields of nanoscience and nanotechnology is well understood. Any development in this filed will have a great deal to do with advances in these techniques. Advances in molecular based study of matter in nanoscale will help to understand, simulate, predict and formulate new materials utilizing the fields of quantum and statistical mechanics, intermolecular interaction, molecular simulation and molecular modeling. We may then be able to understand how to design new molecular building blocks which could allow self-assembly or self-replication to advance the bottom-up approach of producing the necessary materials for the advancement of nanotechnology. The past trend of the contributions of molecular based study of matter in macroscopic technologies is indicative of the fact that its future influence into nanoscience and nanotechnology is quite promising.

Conclusions and Discussion

A momentous scientific and technological activity has begun which is the ability of the human beings to systematically organize and manipulate matter on atomic and molecular level. Significant accomplishments in performance and changes of manufacturing paradigms are predicted to lead to several break troughs in the present, 21st, century.

The answer to the question of how soon will the next industrial revolution arrive depends a great deal on the intensity of activities of the scientific communities in academic disciplines, national laboratories, or even entire industries all around the world. That certainly depends on the efforts by the research and development funding agencies which are mostly powered by government funds. There is also the question of who will benefit the most, and who will be in the position to control and counter the negative aspects of this revolution.

Future developments and implementation of nanotechnology could certainly change the nature of almost every human-made object and activity. Its ultimate societal impact is expected to be as dramatic as the first industrial revolution and greater than the combined influences that aerospace, nuclear energy, transistors, computers, and polymers have had in this century. In the forefront of nanotechnology development is the need to understand the techniques for atomic and molecular based study of matter in nanoscale. To achieve that, the author initiated the writing of the present book. Through techniques of molecular-based study of matter in nanoscale we could contribute more effectively to the advancement of nanotechnology and facilitate the promised next industrial revolution.

Some Important Related INTERNET Sites

Below is a list of the important Internet sites related to nanotechnology:

(i)   APEC on Nanotechnology
      http://www.apectf.nstda.or.th/html/nano.html
(ii) Foresight Institute
      http://www.foresight.org/
(iii) Nanoscale Data and Educational Sites on the Web
      http://www.uic.edu/labs/trl
(iv) U.S. National Nanotechnology Initiative
      http://www.nano.gov

Bibliography

[1]. G. A. Mansoori and F. B. Canfield, *I&EC* **62**, 8,12, (1970).

[2]. J. M. Haile and G. A. Mansoori, *"Molecular Based Study of Fluids"*, Adv. Chem. Series **204**, ACS, Wash., D.C., (1983).

[3]. E. Matteoli and G. A. Mansoori (Ed.s) *"Fluctuation Theory of Mixtures"* Taylor & Francis Publ. Co., New York, NY, (1990).

[4]. E. Keshmirizadeh, H. Modarress, A. Eliassi and G. A. Mansoori, *European Polymer Journal*, **39**, 6, 1141, June (2003).

[5]. NSTC, *"National Nanotechnology Initiative: Leading to the Next Industrial Revolution"* A Report by the Interagency Working Group on Nanoscience", Engineering and Technology Committee on Technology, National Science and Technology Council, Washington, D.C. February (2000).

[6]. M. C. Roco, S. Williams and P. Alivisatos (Editors) *"Nanotechnology Research Directions: IWGN Workshop Report - Vision for Nanotechnology R&D in the Next Decade"* WTEC, Loyola College in Maryland, September (1999).

[7]. A. T. Bell, *Science*, **299**, 1688 (2003).

[8]. W. Heisenberg, *"Physics and Philosophy"*, Harper and Row, New York, NY (1958).

[9]. R. P. Feynman, *"There's plenty of room at the bottom - An invitation to enter a new field of physics"* Engineering and Science Magazine of Cal. Inst. of Tech., **23**, 22, (1960).

[10]. K. E. Drexler, *"Nanosystems: Molecular Machinery, Manufacturing, and Computation"*. John Wiley & Sons, Inc., New York, NY (1992).

[11]. R. D. Piner, J. Zhu, F. Xu, S. Hong, C.A. Mirkin, *Science*, **283**, 661 (1999).

[12]. R. P. Feynman, *"Infinitesimal Machinery"* (1983) - Lecture reprinted in the *Journal of Microelectromechanical Systems*, **2**, 1, 4, March (1993).

[13]. J. von Neumann and A. W. Burks. *"Theory of Self-Reproducing Automata"*, Univ. of Illinois Press, Urbana IL, (1966).

[14]. J. K. Gimzewski, *NovaActa Leopoldina Supplementum Nr.*, **17**, 29, (2001).

[15]. D. M. Eigler and E. K. Schweizer, *Nature* **344**, 524, (1999).

[16]. A. Bezryadin, C. Dekker, and G. Schmid, *Appl. Phys. Lett.* 71, 1273, (1997).

[17]. T. L. Hill. *"Thermodynamics of Small Small Systems"*, I, W.A. Benjamin, New York, NY (1963); *ibid*, II, W.A. Benjamin, New York, NY (1964).

[18]. T. L. Hill, *Nano Letters* 1, 5, 273, (2001).

[19]. G. A. Mansoori, "Nanothermodynamics" in the "Proceedings of the Workshop on NanoSciTech" Kashan University, May 23-24, (2002).

[20]. G. Binnig, H. Rohrer, *Scientific America* 253, 40, Aug. (1985).

[21]. G. Binnig, C. F. Quate, Ch. Gerber, *Phys. Rev. Let.* 56, 930, (1986).

[22]. J. E. Dahl, S. G. Liu, R. M. K. Carlson, *Science* 299, 96 (2003).

[23]. H. Ramezani and G. A. Mansoori, *"Diamondoids as Molecular Building Blocks for Biotechnology (Wet Nanotechnology, Drug Targeting and Gene Delivery)"* (to be published).

[24]. H. W. Kroto, J. R. Heath, S. C. O'Brien, R. F. Curl, and R. E. Smalley, *Nature* 318, 165, (1985).

[25]. W. Krätschmer, L. D. Lamb, K. Fostiropoulos, and D. R. Huffman, *Nature*, 347, 354, (1990).

[26]. A. Eliassi, M. H. Eikani and G. A. Mansoori, *"Production of Single-Walled Carbon Nanotubes"* in *Proceed. of the 1ˢᵗ Conf. on Nanotechnology - The next industrial revolution"*, 2, 160, March (2002).

[27]. S. Iijima, *Nature*, 345, 56, (1991).

[28]. S. Iijima and T. Ichihashi, *Nature*, 363, 603 (1993).

[29]. H. Hesamzadeh, B. Ganjipour, S. Mohajerzadeh, A. Khodadadi, Y. Mortazavi, and S. Kiani, *Carbon* 42, 1043 (2004).

[30]. H. Dai, in *"Carbon Nanotubes, Synthesis, Structure, Properties and Applications"*, M. S. Dresselhaus, G. Dresselhaus, P. Avouris (Eds.), Springer, Heidelberg, p 21, (2001).

[31]. E. T. Mickelson, I. W. Chiang, J. L. Zimmerman, P. J. Boul, J. Lozano, J. Liu, R. E. Smalley, R. H. Hauge and J. L. Margrave (1999), *J. Phys. Chem.*, **B103**, 4318 (1999).

[32]. M. S. Dresselhaus, G. Dresselhaus, and P. C. Eklund, *"Science of Fullerenes and Carbon Nanotubes"*, Academic Press, New York, NY, (1996).

[33]. R. S. Ruoff and D. C. Lorents, *Carbon*, 33, 925, (1995).

[34]. M. F. Yu, B. S. Files, S. Arepalli, and R. S. Ruoff, *Phys. Rev. Lett.*, 84, 5552, (2000).

[35]. R. Andrews, D. Jacques, A. M. Rao, T. Rantell, F. Derbyshire, Y. Chen, J. Chen, and R. Haddon, *Appl. Phys. Lett.*, 75, 1329, (1999).

[36]. D. Tekleab, R. Czerw, P. M. Ajayan and D. L. Carroll, *Appl. Phys. Lett.*, 76, 3594, (2000).

[37]. S. Dag, E. Durgun, and S. Ciraci, *Phys. Rev.* **B 69**, 121407 (2004).

[38]. Y. Gogotsi, J. A. Libera, A. G. Yazicioglu, and C. M. Megaridis, *Applied Phys. Lett.* 79, 7, 1021, 13 Aug. (2001).

[39]. C. M. Megaridis, A. G. Yazicioglu, and J. A. Libera and Y. Gogotsi, *Phys. Fluids*, 14, L5, 2 Feb. (2002).

[40]. J. M. Frechet and D. A. Tomalia *"Dendrimers and Other Dendritic Polymers"* John Wiley & Sons, Inc., New York, NY, (2001).

[41]. J. M. Alexander, J. L. Clark, T. J. Brett, and J. J. Stezowski, PNAS, **99**, 5115 (2002).

[42]. S.-S. Feng, G. Ruanb and Q.-T. Lic, *Biomaterials,* **25**, 21, 5181, September (2004).

[43]. N. E. Thompson, K. M. Foley and R. R. Burgess, *Protein Expression and Purification,* **36**, 2, 186, August (2004).

[44]. S. Priyanto, G. A. Mansoori and A. Suwono, *Chem. Eng. Science* **56**, 6933, (2001).

[45]. G. A. Mansoori, "Organic Nanostructures and Their Phase Transitions" in *Proceed. of the 1ˢᵗ Conf. on Nanotechnology - The next industrial revolution"*, **2**, 345, March (2002).

[46]. H. Rafii-Tabar and G. A. Mansoori, Encycl. of Nanosci. & Nanotech., **4**, 231, (2004).

[47]. P. Preuss, Science Beat: *http://www.lbl.gov/Science-Articles/Archive/ALS-EUVL-sidebar.html)*, July 14, (2004).

[48]. I. Takashi and O. Shinji, *Nature,* **406**, 1027 (2000).

Chapter 2

Nanosystems Intermolecular Forces and Potentials

"It will be perfectly clear that in all my studies I was quite convinced of the real existence of molecules, that I never regarded them as a figment of my imagination, nor even as mere centres of force effects." **J. D. van der Waals**

Introduction

In this chapter, we present details of interatomic and intermolecular forces and potential energy functions from the point of view of formation and functions of nanostructures. Detailed information about interatomic and intermolecular interactions is necessary for modeling, prediction and simulation of the behavior of assembly of finite number of molecules, which happen to exist in nanosystems. In addition such information will guide the design of positional-assembly and prediction of self-assembly and self-replication which are the fundamentals of bottom-up nanotechnology.

In this chapter, we first introduce a short description of covalent interactions and their differences with non-covalent interactions. Then we will present a detailed analysis of the non-covalent and covalent interactions. This will include experimental and theoretical modeling followed by phenomenological models developed for non-covalent and covalent interactions.

Covalent and Noncovalent Interactions

The interactions between atoms and molecules are either of covalent or non-covalent type [1,2]. Covalent interactions are in a form of chemical bond in compounds that result from the sharing of one or more pairs of electrons. Atoms can combine by forming molecules to achieve an octet of valence electrons by sharing electrons. In this process, the electron clouds of every two atoms overlap where they are thicker and their electric charge is stronger. Since the nuclei of the two atoms have stronger attractive potentials to their respective thick electron clouds than their mutual repulsive potential (due to their far distance from one another), the two atoms are held together and form a molecule (see Figure 1).

Figure 1. Symbolic demonstration of covalent bond formation between two atoms.

As a result, the formation of covalent bonds requires overlapping of partially occupied orbitals of interacting atoms, which share a pair of electrons. On the other hand, in non-covalent interactions no overlapping is necessary because the attraction comes from the electrical properties of the interacting atoms and molecules. The electrons in the outermost shell are named the **valence electrons**, which are the electrons on an atom that can be gained or lost in a chemical reaction. The number of covalent bonds, which an atom can form, depends on how many valence electrons it has. Generally, covalent bond implies the sharing of just a single pair of electrons. The sharing of two pairs of electrons is called a double bond and three pairs sharing is called a triple bond. The latter case is relatively rare in nature, and actually two atoms are not observed to bond more than triply. Most frequently covalent bonding occurs between atoms, which possess similar electronegativities. That is when they have similar affinities for attracting electrons. This occurs when neither atom can possess sufficient affinity, or attractive potential energy, for an electron to completely remove an electron from the other atom.

Covalent interactions are stronger than the non-covalent bonds. The weak non-covalent interactions were first recognized by J. D. van der Waals in the nineteenth century [2, 3]. Covalent interactions are of short range and the resulting bonds are generally less than 0.2 ϕ [nm] long. The non-covalent interactions are of longer range and within a few ϕ [nm] range. In Table 1 the differences between covalent and non-covalent interactions are reported.

Table 1. Differences between covalent and non-covalent interactions.

INTERACTION TYPE	COVALENT	NON-COVALENT
BOND TYPE	Chemical	van der Waals, hydrogen bonding, etc.
ENERGETIC (kcal/mol)	25-200	0.1-5
STABILITY	Stable	Open to change
CONTRIBUTION TO ΔG	Mainly ΔH	Both ΔH and TΔS
NATURE	Chemical reaction	Chemical equilibrium
MEDIUM/SOLVENT EFFECTS	Secondary	Primary
COOPERATIVE BONDING	Less important	Very important

Interatomic and Intermolecular Potential Energies and Forces

In order to have a successful theoretical model or simulation, one needs to adopt a physically correct inter-particle potential energy model. The majority of the interatomic and intermolecular potential energy models, e.g., Lennard-Jones, $\phi(r)=4\varepsilon[(\sigma/r)^{12}-(\sigma/r)^{6}]$, are designed to give a statistically-averaged (effective) representation of such forces for macroscopic systems consisting of many particles. Even the ranges of accuracy of the available interparticle potential energy parameters and constants are limited and are aimed at the prediction of certain macroscopic properties. That is probably why the application of the existing interatomic and intermolecular force models for prediction of nano-crystalline structures, fullerene, nanotube, diamondoids, aggregation of amphiphilic molecules into micelles and coacervates, biological macromolecules interactions, such as between DNA and other molecules, are not quantitatively accurate. The interatomic and intermolecular potential energy database for rather simple fluids and solids in macroscale are rather complete. Parameters of interaction energies between atoms and simple molecules have been calculated through such measurements as x-ray crystallography, light scattering, nuclear magnetic resonance spectroscopy, gas viscosity, thermal conductivity, diffusivity and the virial coefficients data. Most of the present phenomenological models for interparticle forces are tuned specifically for statistical mechanical treatment of macroscopic systems. However, such information may not be sufficiently accurate in treatment of nanosystems where the number of particles is finite and validity of the existing statistical averaging techniques fail [4-11].

Exact knowledge about the nature and magnitude of interatomic and intermolecular interactions is quite important for the correct prediction of the behavior of nanostructures. For example, in the process of formation of crystalline nanostructures close packing is achieved when each particle is touching six others arranged so that their centers form a hexagon (see Figure 2).

Figure 2. Three dimensional close-packing options due to differences in intermolecular potential energies.

Three dimensional close-packing can be created by adding layers of close-packed planes one above the other so that each particle is in contact with three others in each of the planes directly above and below it. There are two mutually exclusive options for the positions of the particles when commencing a new layer, as indicated by the black and white interstices of Figure 2. One may associate these options to interatomic and intermolecular potential energy differences. This is an example of the role of such forces in structure. Knowledge about ineratomic and intermolecular interactions is very critical in designing and fabricating nanostructure-based systems and devices such as phase transitions, self-assemblies, thin film transistors, light-emitting devices, semiconductor nanorods, composites, etc. However, the nature and role of intermolecular interactions in nanostructures is very challenging and not well-understood [6-10].

Intermolecular potential energies include pairwise additive energies, as well as many-body interactions. The interparticle interaction potential energy between atoms and molecules is generally denoted by $\Phi(r)=\Phi_{rep}+\Phi_{att}$, where r is the intermolecular distance, Φ_{rep} is the repulsive interaction energy and Φ_{att} is the attractive interaction energy, see Figure 3. From the equation above, the interaction force is $F = - [\partial\phi(r)/\partial r]_{\theta i\theta j,\varphi} = F_{rep} + F_{att}$.

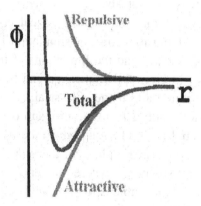

Figure 3. The pair interaction energies.

Among pairwise additive energies one can mention the repulsive potentials, van der Waals energies, interactions involving polar and polarization of molecules, interactions involving hydrogen bonding and coulomb interactions. Among many-body interactions one can name the Axilrod-Teller triple-dipole interactions [12,13].

Molecular structural characteristics of macroscopic systems based on the knowledge of statistically averaged (effective) pair-interparticle interactions are already defined and formulated in the thermodynamic limit [12-16]. As a result, the available atomic and molecular interaction database for macrobodies by means of effective-pair-intermolecular potentials is only an approximation. Since the number of particles is finite in nanostructures, those statistically averaging tools are not well suited for this purpose. As a result, the theory of intermolecular potential functions of such systems and their relationship with their structural characteristics need to be formulated.

For atoms and rather simple molecules, quantum mechanical *ab initio* calculation methods [17-19] have been successful in producing accurate intermolecular potential functions. For complex molecules and macromolecules, the computation is prohibitively difficult to produce accurate potential data.

Classical potentials are not always very accurate, and they usually have a simple analytical form so that numerical computations can be made fast. So, when doing modeling or simulation [20], one should be perfectly aware of the power and the limitations of the potential models being used. One should know in what range and for what purposes the potential model was fitted and made. Classical, and even tight-binding, potential models always suffer from the problem of transferability. For example if the potential is fitted to reproduce the mechanical properties of the face-centered cubic (FCC) phase of a crystal, one usually can not expect it to work in situations where the environment of atoms is different, like in a simple cubic structure with six nearest neighbors instead of twelve in FCC. Or one can not expect to obtain correct defect energies if the potential model is only fitted to the elastic properties of the bulk phase [21-24]. The analytic forms of those modern inter-particle potentials are usually not so simple. While their two-body interaction forms are complicated they could have more than pair interactions in them. Three-body or four-body terms may be added to the two-body part. In some other instances, the potential has a two-body form with parameters that now have become dependent on the number and distances of the neighboring atoms.

Now that we have emphasized the importance and limitations of the interatomic and intermolecular potentials in theoretical modeling and simulation, we will also discuss the general forms adopted for these potentials and the way they are determined. Inter-particle forces and potential energies are known to be, generally, orientation-dependent and can be written as a sum of repulsive, London dispersion, hydrogen bonding and electrostatic energies, i.e.

$$\phi\left(r, \theta_i, \theta_j, \varphi\right) = \phi_{rep} + \phi_{disp} + \phi_{dipole} + \phi_{quad.} + \phi_{ind.dipole} + \ldots, \quad (1)$$

where r is the separation of the two bodies, θ_i, and θ_j are the angles between the molecule axes and the bonds between molecules, and φ is the azimuthal angle. For neutral and spherically symmetric molecules when the separation r is very small, an exponential repulsive term,

$$\phi_{rep} = \alpha \cdot exp(\beta/r), \quad (2)$$

dominates and the potential is strongly positive. Hence, ϕ_{rep} describes the short-range repulsive potential due to the distortion of the electron clouds at small separations. For neutral and spherically-symmetric molecules when the separation r is large, the London dispersion forces dominate.

On the experimental front, the most significant developments were brought about by the invention of the scanning tunneling microscope (STM) in 1982 [25], followed by the atomic force microscope (AFM) [26] in 1986. These are tip-based devices, which allow for a nanoscale manipulation of the morphology of the condensed phases and the determination of their electronic structures. These probe-based techniques have been extended further and are now collectively referred to as the scanning probe microscopy (SPM). The SPM-based techniques have been improved considerably, providing new tools in research in such fields of nanotechnology as nanomechanics, nanoelectronics, nanomagnetism and nanooptics [27].

The use of AFM and its modification to optical detection [28] has opened new perspectives for direct determination of interatomic and intermolecular forces. For atoms and molecules consisting of up to ten atoms, quantum mechanical *ab initio* computations are successful in producing rather exact force-distance results for inter-particle potential energy [18,29-32]. For complex molecules and macromolecules one may produce the needed intermolecular potential energy functions directly through the use of atomic force microscope (AFM). For example, atomic force microscopy data are often used to develop accurate potential models to describe the intermolecular interactions in the condensed phases of such molecules as C_{60} [31], colloids [34], biological macromolecules [35], etc. The non-contact AFM can be used for attractive interactions force measurement. Contact AFM can be used for repulsive force measurement. Intermittent-contact AFM is more effective than non-contact AFM for imaging larger scan sizes. Because of the possibility to use the AFM in liquid environments [36,37] it has become possible to image organic micelles, colloids, biological surfaces such as cells and protein layers and generally organic nanostructures [38] at nanoscale resolution under physiological conditions.

Also making microelectrophoretic measurements of zeta potential allows us to calculate the total interparticle energies indirectly [39]. From the combined AFM and microelectrophoretic measurements accurate force-distance data could be obtained [40]. From the relation between the force and distance, an interparticle force vs. distance curve can be created. Then with the use of the phenomenological potential functions [10] the produced data can be readily fitted to a potential energy function for application in various nanotechnology and nanoscience computational schemes.

Experimental and Theoretical Development of Interparticle Potentials

Theoretically, there are tremendous challenges because there is no simple way to extract potential energy surfaces from force measurements. A given potential energy surface is a function of the many coordinates of the nanoparticles. Particularly for large molecules, there might be many different potential surfaces corresponding to the same force measurement. Nevertheless, one may be able to tackle this problem with a three-pronged approach:

(1): Based on modeling experimental data, one may proceed with a top-down strategy. Namely, starting from the experimental data, one can construct a model for different molecules and clusters while drawing on empirical methods to help extract the essential features and build an appropriate potential model for a given compound.

(2): One may also explore a bottom-up approach, by using first-principles methods (density functional theory [41] and quantum chemical techniques [42]) to calculate the potential surfaces for simple molecules. Here we are virtually free of any input parameters by constructing the model potentials and testing them on much larger molecules. This step can be iterated many times by comparing the quality of the potential with *ab initio* calculations. Once a reliable potential is obtained, it may be used to compute the properties associated with different structures in real materials. At this stage, almost no *ab initio* calculations can be done practically since they would exhaust the present available computational power. This readily leads us to the third step.

(3): One may directly compare theoretical predictions with the experimental models obtained from (**1**). This is central to a project since from the comparison, it will become possible to update the experimental models and initiate new developments with *ab initio* calculations. Once good consistency is found, it should be possible to predict and design new materials. Figure 4 shows how these studies are linked together, and details are presented following the figure.

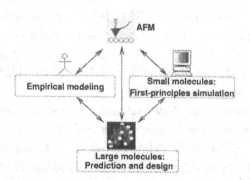

Figure 4. Flowchart for the experimental plus theoretical interparticle force study. Starting from the AFM results, an empirical model could be constructed. Both first-principles calculations and experimental results could then be used to predict the properties of materials and ultimately to design new materials.

Step (1): AFM Measurement and Empirical Modeling

The invention of the atomic force microscope, AFM, [26] in 1986 and its modification to optical detection [28] has opened new perspectives for various micro- and nanoscale surface imaging in science and industry. The use of AFM not only allows for nanoscale manipulation of the morphology of various condensed phases and the determination of their electronic structures; it can be also used for direct determination of interatomic and intermolecular forces.

However, its use for measurement of interparticle interaction energies as a function of distance is getting more attention due to various reasons. For atoms and molecules consisting of up to ten atoms, quantum mechanical *ab initio* computations are successful in producing rather exact force-distance results for interparticle potential energy. For complex molecules and macromolecules one may produce the needed intermolecular potential energy functions directly only through the use of atomic force microscope (AFM). For example, atomic force microscopy data are often used to develop accurate potential models to describe the intermolecular interactions in the condensed phases of such molecules as C_{60} [33].

The atomic force microscope (AFM) is a unique tool for direct study of intermolecular forces. Unlike traditional microscopes, AFM does not use optical lenses and therefore it provides very high-resolution range of various sample properties [26,43]. It operates by scanning a very sharp tip across a sample, which 'feels' the contours of the surface in a manner similar to the stylus tracing across the grooves of a record. In this way it can follow the contours of the surface and so create a topographic image, often with sub-nanometer resolution.

This instrument also allows researchers to obtain information about the specific forces between and within molecules on the surface. The AFM, by its very nature, is extremely sensitive to intermolecular forces and has the ability to measure force as a function of distance. In fact measurement of interactions as small as a single hydrogen bond have been reported [34,44-46]. The non-contact AFM will be used for attractive interactions force measurement. Contact AFM will be used for repulsive force measurement. Intermittent-contact AFM is more effective than non-contact AFM for imaging larger scan sizes.

In principle to do such a measurement and study with AFM it is necessary to specially design the tip for this purpose [28,47,48]. Sarid [27] has proposed force-distance relationships when the tip is made of a molecule, a sphere, and a cylinder assuming van der Waals dispersion attractive forces. Various other investigators have developed the methodologies for force-distance relationship for other tip geometric shapes including cylinder, paraboloid, cone, pyramid, a conical part covered by the spherical cap, etc [43,47-55]. For example, Zanette et al [47] present a theoretical and experimental investigation of the force-distance relation in the case of a pyramidal tip. Data analysis of interaction forces measured with the atomic force microscope is quite important [56]. Experimental recordings of direct tip-sample interaction can be obtained as described in [57] and recordings using flexible cross-linkers can be obtained as described in [58,59]. The noise in the typical force-distance cycles can be assumed to be, for example, Gaussian. Recent progress in AFM technology will allow the force-distance relationship measurement of inter- and intra-molecular forces at the level of individual molecules of almost any size.

Because of the possibility to use the AFM in liquid environments [36,37] it has become possible to image organic micelles, colloids, biological surfaces such as cells and protein layers and generally organic nanostructures [38] at ϕ [nm]-resolution under physiological conditions. One important precaution in the force measurement is how to fix micelles, colloids, biological cells on a substrate and a probe, secure enough for measuring force but flexible enough to keep the organic nanostructure intact. In the case of biological cells one must also keep it biologically active [37]. Variety of techniques for this purpose have been proposed including the use of chemical cross-linkers, flexible spacer molecules [60], inactive proteins as cushions in case of biological systems [61] and self-assembled monolayers [62]. In liquid state force-distance measurements an important issue to consider is the effect of pushing the organic nanostructures on the substrate and AFM probe. As the AFM probe is pushed onto the nanostructure, there is a possibility of damaging it or physically adsorbing it to the probe.

Also making microelectrophoretic measurements of zeta potential will allow us to calculate the total interparticle energies indirectly. From the combined AFM and microelectrophoretic measurements accurate force-distance data can be obtained. From the relation between the force and distance, an interparticle force vs. distance curve can be created. Then with the use of the phenomenological potential functions presented in this review the produced data can be readily fitted to a potential energy function for application in various nano computational schemes.

Atomic force microscopy can provide a highly-accurate measure of particle-particle forces by scanning a very sharp tip across a sample. This force depends on the surface morphology and local type of binding – covalent and non-covalent (van der Waals, hydrogen bonding, ionic, etc.). The change of force vs. distance reflects the nature of the potential between the molecules on the surface and the molecules on the tip. There are two tasks to be performed at this step:

(1) Since the electronic, Coulomb and exchange interactions are functions of the inverse distance between electrons, normally the variation of potential vs. distance is nonlinear. The information from such experimental relation between force and distance will be used to construct a model potential,

$$\phi(r_{i,j}) = \sum_{i,j} c_{i,j} f(r_{i,j}), \tag{3}$$

where $c_{i,j}$'s are the fitting parameters and $r_{i,j}$'s represent the coordinates of the atoms. In general, f should be a function of both distance and angle. If we assume that f is a function of distance, there are several commonly-used functional forms for f in the literature as shown in Table 2.

(2) The potential function also includes higher-order interactions (three-body, four-body, etc.). Since normally the potential is also a function of angle bend, stretch-bend, torsion and rotation, one may perform angular-dependent force measurements. This would provide detailed information about the interactions between different molecules at different angles.

Table 2. Some pair potential functions.

Name	Form	Applicable Range
Harmonic oscillator	$(^1/_2)K(r_1 - r_2)^2$	Intermediate (chemically bonded)
Morse	$D[e^{-2a(r_1-r_2)} - 2e^{-b(r_1-r_2)}]$	Intermediate (chemically bonded)
Mie[(*)]	$[\dfrac{m}{(m-n)}](\dfrac{m}{n})^{\frac{n}{(m-n)}} \varepsilon[(\dfrac{\sigma}{r})^m - (\dfrac{\sigma}{r})^n]$	Long-range (non-bonded)
Coulomb	$\dfrac{q_1 q_2}{r}$	Ionic bonding (partial or full electron transfer)

(*) Lennard-Jones is the special case of Mie when m=12, n=6.

After finishing these two tasks, it should be possible to build sophisticated potential surfaces which are functions of, both, distance and angles. These potentials could then be used to construct intramolecular and intermolecular potentials. Both of them are of critical importance for understanding and designing, for example, biochemical molecules.

The obtained potentials could be used to calculate the images of the nanostructures under investigation in different AFM scanning regimes (analogous to those shown in Figure 5). Such theoretical images could be verified with the highly accurate AFM measurements.

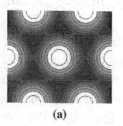
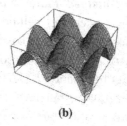

(a)   **(b)**

Figure 5. Theoretical simulation of the scanning process at a constant height in the AFM repulsive mode above the closely-packed surface. Both (a) 2-d and (b) 3-d images are shown for the vertical force component acting on the AFM tip depending on its position above a surface. In (a) the darker tone denotes weaker force and the lighter tone the stronger force. Analogous but more complicated figures can be obtained for any model potential describing the intermolecular interaction. Comparison of such kind of theoretical figures with the experimental ones will help to make a choice of the most adequate model of the potential.

By using potentials that have several levels of accuracy in a computer simulation calculation or a molecular-based theoretical model, one must know what kind of result to expect and always keep in mind for what kind of calculations these potentials can be used [20]. For instance, if a potential is fitted to reproduce elastic properties of the diamond phase of carbon, then one will use this potential to perform simulation or calculation of the diamond phase and can not expect in general, to get accurate answers for the elastic or other properties of another phase of carbon with a different coordination number.

Step (2): Theoretical Modeling

One may use the full-potential linearized augmented plane wave (FLAPW) method within the density functional theory (WIEN package [63] and an appropriate pseudopotential code [64]) along with the available quantum chemical techniques (Gaussian98 [17] and MOLPRO [18]) to compute the potential surfaces for a sample molecule. In the following, we briefly describe the FLAPW method, which is among the most accurate density functional based methods, while a full description can be found in [43].

Linearized Augmented Plane Wave (LAPW): The LAPW method is among the most accurate methods for performing electronic structure calculations for crystals. It is based on the density functional theory for the treatment of exchange and correlation, and utilizes, e.g., the local spin density approximation (LSDA). Several forms of LSDA potentials exist in the literature, but recent improvements using the generalized gradient approximation are available as well. For valence states, relativistic effects can be included either in a scalar relativistic treatment or with the second variational method including spin-orbit coupling. Core states are treated fully relativistically. This adaptation is achieved by dividing the unit cell into (I) non-overlapping atomic spheres (centered at the atomic sites) and (II) an interstitial region (Figure 6).

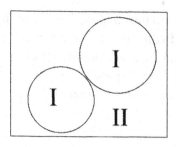

Figure 6. Partitioning of the unit cell into atomic spheres (I) and an interstitial region (II).

In the two types of regions, different basis sets are used:

(I) Inside an atomic sphere of radius R_{MT} (where MT refers to "muffin tin" sphere), a linear combination of radial functions times spherical harmonics $Y_{lm}(\hat{r})$ is used,

$$\phi_{k_n} = \sum_{lm} [A_{lm} u_l(r, E_l) + B_{lm} \dot{u}_l(r, E_l)] Y_{lm}(\hat{r}), \tag{4}$$

where $u_l(r,E_l)$ is (at the origin) the regular solution of the radial Schrödinger equation for the energy E_l (chosen normally at the center of the corresponding band with l-like character), and the spherical part of the potential inside the sphere is the energy derivative of u_l (i.e., $\dot{u}$) taken at the same energy E_l. A linear combination of these two functions constitutes the linearization of the radial function. The coefficients A_{lm} and B_{lm} are functions of k_n (see below) determined by requiring that this basis function matches (in value and slope) the corresponding basis function of the interstitial region; these are obtained by numerical integration of the radial Schrödinger equation on a radial mesh inside the sphere.

(II) In the interstitial region, a plane wave expansion is used,

$$\phi_{k_n} = \frac{1}{\sqrt{\Omega}} e^{ik_n r}, \tag{5}$$

where $k_n = k + K_n$, K_n are the reciprocal lattice vectors, and k is the wave vector inside the first Brillouin zone. Each plane wave is augmented by an atomic-like function in every atomic sphere. The solutions to the Kohn-Sham equations are expanded in this combined basis set of LAPWs according to the linear variation method,

$$\psi_k = \sum_n c_n \phi_{k_n}, \qquad (6)$$

and the coefficients c_n are determined by the Rayleigh-Ritz variational principle. The convergence of this basis set is controlled by a cutoff parameter $R_{MT} K_{max} = 6 - 9$, where R_{MT} is the smallest atomic sphere radius in the unit cell, and K_{max} is the magnitude of the largest K vector.

Full-Potential Linearized Augmented Plane Wave (FLAPW): In its most general form, the LAPW method expands the potential in the form,

$$\phi(R) = \sum_{lm} \phi_{lm}(r)\, Y_{lm}(\hat{r}) \text{ for inside the sphere and } \phi(R) = \sum_K \phi_K e^{iKr}, \qquad (7)$$

for outside the sphere, and the charge densities analogously. Thus, no shape approximations are made. This procedure is frequently called the "full-potential" LAPW method, i.e., FLAPW. The forces at the atoms are calculated according to Ru et al. [65]. For implementation of this formalism with the WIEN package, see Kohler et al. [66]. An alternative formulation by Soler and Williams [67] has also been tested and found to be equivalent [68].

The advantage of these theories is that they are normally parameter-free. This is usually an independent check of experimental results and the model potentials. Moreover, with more than fifty years of development, these theories have become quite sophisticated and have strong reliability and predictability. They essentially can treat all kinds of atoms and molecules. For instance, recently *ab initio* theory is used to simulate the reaction path for a retinal molecule in rhodopsin [30-32] (see Figure 7).

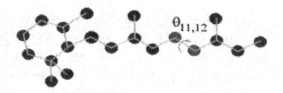

Figure 7. Rhodopsin segment. Isomerization occurs around the bond 11-12 with $\theta_{11,12}$ angle.

It is shown that not only the ground state, Figure 8(a), but also the excited states, Figure 8(b), can be calculated precisely.

It is well known that the potential resulting from electrons is much more difficult to treat than that from atoms; this is because the electron density is extremely sensitive to the structure change and

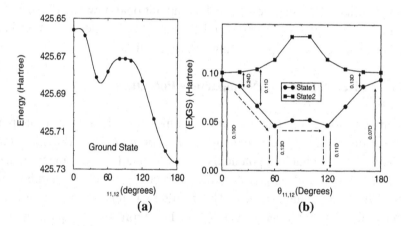

Figure 8. (a) Ground-state potential versus angle $\theta_{11,12}$ in a segment of rhodopsin. (b) Potential for the excited state versus the angle around the bond 11-12. Importantly, the transition matrix elements are also accurately obtained [25-28].

external perturbations. Experimentally, the electronic contribution is often very difficult to probe since an external intervention will affect the intrinsic electron distribution substantially. Theoretically, one has the flexibility to calculate these intrinsic properties. One may start from small molecules where no difficulty is expected. This way the potential should be computed easily and is the microscopic potential between molecules and atoms. Building on this effort, one may then calculate larger molecules and clusters. This is a very important leap, where two levels of calculations could be done: (a) from the above microscopic potential, it is then possible to construct a model potential for larger clusters and nanoparticles, and (b) massive parallel computations of realistic potentials for larger nanoparticles consisting of many atoms could be carried out. One may exploit the advantage of parallel computing to find the potential. In particular, for nanoparticles, one can use the chemist's view to treat different atoms by different computers or nodes. Interactions between these nodes could be communicated through fast Internet connections. With the increase of computational power, one is able to compute even larger clusters, where the potential is "exact". The model potential resulting from (a) can be directly compared with this potential by improving the quality of the model potential and predictability. This step needs much computational power and is usually iterated many times until one could find a reliable potential. This is a very critical step before proceeding to Step 3 below.

Step (3): Development of Nanoparticle Potentials

When a cluster or nanoparticle becomes larger, a first-principles calculation becomes very expensive and prohibitive. Provided one already has reliable potentials from small and medium-size nanoparticles, it will be possible to construct independent potentials for large cluster or nanoparticle. The resultant potential can be directly compared with empirical models (Step 1). From the comparison, one can refine the experimental models. Such comparison could be repeated until the model potential is optimized. Another direct consequence from this comparison is that it helps one to identify some areas for further theoretical developments. Experimentally, features such as impurities, structural disorders, grain boundaries and oxidation may be present in real samples. Theoretically, depending on the level of sophistication – self-consistent field, configuration interaction, multi-reference self-consistent field and multi-reference configuration interaction – the present first-principles calculations may give different potential surfaces. This is especially true when the compounds have transition metal elements, as often is the case in biological molecules, where the electron correlation is strong. Here, perturbation theory breaks down, although one may be able to develop a very accurate density-matrix renormalization group technique for model systems [30-32]. The initial results are quite promising, where one sees a clear improvement over traditional quantum chemistry methods. The main idea of this method is to retain the most relevant information while properly treating the boundary condition. This method is probably one of the most accurate methods for treating large systems. With the improved quality of theoretical calculations and model simulations, one should be ready to predict and design new materials with desirable features.

Phenomenological Interatomic and Intermolecular Potentials

The procedure and steps presented above is important when dealing with nanosystems composed of up to a few hundred atoms and molecules, or macro- and supra-molecules for which potential functions are not available. To study nanostructures composed of several hundred to several million atoms or molecules including macroscopic systems, the computationally most efficient method is the use of phenomenological interatomic and intermolecular potentials.

The phenomenological potentials are obtained by using empirical approaches of selecting a mathematical function and fitting its unknown parameters to various, experimentally determined, properties of the system, such as its lattice constant.

Interatomic and intermolecular potentials must be able to model the energetics and dynamics of nanostructures, and this fact lies at the very foundation of the computer-based modeling and simulations. The significance of much of the modeling and simulation results, their accuracy and the extent to which they represent the real behavior of nanostructures, and their transitions, under varied conditions, depends in a critical manner on the accuracy of the interatomic and intermolecular potentials employed.

A great deal of effort has been spent over the years to develop phenomenological intermolecular potentials to model the bonding in various classes of materials, such as metallic, semi-metallic, semi-conducting, and organic atoms and molecules. For a review see [10,69-71].

To be effective for computational nanotechnology, interatomic and intermolecular potentials must possess the following properties [10,72,73]:

(a) Flexibility: A potential energy function must be sufficiently flexible that it could accommodate as wide a range as possible of fitting data. For solid systems, this data might include lattice constants, cohesive energies, elastic properties, vacancy formation energies, and surface energies.

(b) Accuracy: A potential function should be able to accurately reproduce an appropriate fitting database.

(c) Transferability: A potential function should be able to describe at least qualitatively, if not with quantitative accuracy, structures not included in a fitting database.

(d) Computational efficiency: Evaluation of the function should be relatively efficient depending on quantities such as system sizes and time-scales of interest, as well as available computing resources.

In this section we shall describe some of the potential functions that meet these criteria, and are widely used in nano computations.

1. Interatomic Potentials for Metallic Systems

Bonding in metallic systems operates within the range of 0.2 to 0.5 ϕ [nm] [10,24]. At large interatomic distances, the predominant forces arise from van der Waals interactions, which are responsible for long-range cohesion. Metallic bonding, like covalent bonding, arises from the sharing of electrons and hence its proper description requires the consideration of the many-body effects. Two-body potentials are incapable of describing this bonding [74,75] since:

(a) For most cubic metals, the ratio of the elastic constants, C_{12} to C_{44}, is far from unity, whereas a pairwise potential leads to the Cauchy relation, i.e. $C_{12} = C_{44}$.

(b) The prediction of the unrelaxed vacancy formation energy gives values around the cohesive energy which is completely incorrect for metals. The relaxation energy for metals is quite small and the experimental data suggest that the vacancy formation energy for metals is about one third of the cohesive energy.

(c) The interatomic distance between the first and second atomic layers within an unreconstructed surface structure (bulk cross section) is predicted to be expanded by pairwise potentials. This is in contrast with the experimental data which suggest a contraction of the open surface lattice spacing, i.e. pair potentials fail to predict an inward relaxation of the metallic surfaces.

(d) Pairwise potentials overestimate the melting point by up to 20% of the experimental value.

(e) Potentials with a functional form having only one optimum at the diatomic equilibrium distance cannot be fitted properly to the phonon frequencies. Two approaches have been proposed for going beyond pair potentials and incorporating many-body effects into two-body potentials.

The first approach is to add a term, which is a functional of the local electronic density of a given atom, to the pairwise term. This method by itself has led to several alternative potentials that mimic the many-body effects. These many-body potentials are known as the embedded-atom model (EAM) potentials [76-78], which have been employed in several studies involving elemental metals and their alloys [79-84], the Glue Model potentials [85], the Finnis-Sinclair potentials for the BCC elemental metals [86], which have also been developed for the noble metals [87], the Sutton-Chen (SC) potentials [88] for the ten FCC elemental metals, and the Rafii-Tabar and Sutton potentials [89] for the FCC random binary alloys which have also been used in several modeling studies [69,90-92].

The second approach is to go from pair potentials to cluster potentials by the addition of higher order interactions, for example three-body and four-body terms, with appropriate functional forms and symmetries. This has led to potentials, such as the Murrell-Mottram cluster potentials [75]. Inclusion of higher-order terms provides a more accurate modeling of the energetic of the phenomena than is given by pair potentials alone. In the following sections, we consider the potentials pertinent to each approach.

1.1. The Many-Body Embedded-Atom Model (EAM) Potentials: The many-body EAM potentials were proposed [76-78] to model the bonding in metallic clusters. They were the first alternatives to the traditional pair potential models. Their construction is based on the use of density functional theory (DFT), according to which the energy of a collection of atoms can be expressed exactly by a functional of its electronic density [93]. Similarly, the energy change associated with embedding an atom into a host background of atoms is a functional of the electronic density of the host before the new atom is embedded [94,95]. If we can find a good approximation to the embedding functional, then an approximate expression for the energy of an atom in a metal can be constructed.

The total electron density of the host atoms is approximated as a linear superposition of the electron densities (charge distributions) of individual host atoms. To zeroth order, the embedding energy can be equated to the energy of embedding an atom in a homogenous electron gas, whose density, $\rho_{h,i}$, matches the host density at the position of the embedded atom, augmented by the classical electrostatic interaction with the atoms in the host system [96]. The embedding energy for the homogeneous electron gas can be calculated from an *ab initio* basis. Computation of $\rho_{h,i}$ from a weighted average of the host density over the spatial extent of the embedded atom improves the description by accounting for the local inhomogeneity of the host density. The classical electrostatic interaction reduces to a pairwise sum if a frozen atomic charge density is assumed for each host atom [96]. This approach, called quasi-atom method [94], or the effective-medium theory [95], provides the theoretical basis of the EAM, and similar methods.

In the EAM model, the total energy of an elemental system is, therefore, written as

$$\Phi_I^{EAM} = \sum_i F_i \left[\rho_{h,i}\right] + (\tfrac{1}{2})\sum_i\sum_{j\neq i} \phi_{ij}\left(r_{ij}\right), \tag{8}$$

where $\rho_{h,i}$ is electron density of the host at the site of atom i, $F_i[\rho]$ is the embedding functional, *i.e.* the energy to embed the atom i into the background electron density, ρ, and ϕ_{ij} is a pairwise central potential between atoms i and j, separated by a distance r_{ij}, and represents the repulsive core-core electrostatic interaction. The host electron density is a linear superposition of the individual contributions, and is given by,

$$\rho_{h,i} = \Sigma_{j \neq i} \, \rho_j^*(r_{ij}) \,, \qquad (9)$$

where ρ_j^*, another pairwise term, is the electron density of atom j as a function of interatomic separation. It is important to note that the embedding functional, $F_i[\rho]$, is a universal functional that does not depend on the source of the background electron density. This implies that the same functional is employed to compute the energy of an atom in an alloy as that employed for the same atom in a pure elemental metal [79]. Indeed, this is one of the attractive features of these potentials. For a solid at equilibrium, the force to expand, or contract, due to the embedding function is exactly balanced by the force to contract, or expand due to the pairwise interactions. At a defect, this balance is disrupted, leading to the displacements as atoms move to find a new balance [96]. The positive curvature of F plays a key role in this process, by defining the optimum trade off between the number of bonds and the length of those bonds.

The expression for the Cauchy pressure for a cubic crystal can be found from Eq. (8), and is seen to depend directly on the curvature of the function F as described in [75],

$$C_{11} - C_{44} = (1/\Omega)(d^2F/d\rho_{h,i}^2)[\Sigma_j \, (d\rho/dr_{ij})(x_{ij}^2/r_{ij})]^2 \,, \qquad (10)$$

where Ω is the atomic volume and x_{ij} is the x-component of the r_{ij}. To apply these potentials, the input parameters required are the equilibrium atomic volume, the cohesive energy, the bulk modulus, the lattice structure, as well as the repulsive pair potentials and the electron density function [81]. Among the extensive applications of these potentials, we can list their parameterization and use in the computation of the surface energy and relaxation of various crystal surfaces of *Ni* and *Pd* and the migration of hydrogen impurity in the bulk *Ni* and *Pd* [77], the computation of the formation energy, migration energy of vacancies and surface energies of a variety of FCC metals [79], the calculation of the surface composition of the *Ni-Cu* alloys [97], the computation of the elastic constants and vibrational modes of the Ni_3 *Al* alloy [80], the self-diffusion and impurity diffusion of the FCC metals [82], the computation of the heats of solution for alloys of a set of FCC metals [83], and the computation of the phase stability of FCC alloys [84]. There has also been an application of these potentials to covalent materials, such as *Si* [98].

In a rather recent application [99], the second-order elastic moduli (C_{11}, C_{12}, C_{44}) and the third-order elastic moduli (C_{111}, C_{112}, C_{123}, C_{144}, C_{166}, C_{456}), as well as the cohesive energies and lattice constants, of a set of 12 cubic metals with FCC and BCC structures were used as input to obtain the corresponding potential parameters for these metals [100]. The resulting potentials were then used to compute the pressure-volume (P - V) curves, phase stabilities and the phonon frequency spectra, with excellent agreement obtained for the P - V curves with the experimental data, and a reasonable agreement obtained for the frequency curves.

The EAM potentials can also be written for ordered binary alloys [96]. One can write

$$\Phi_{Alloy}{}^{EAM} = \sum_i F_{ti} \left[\rho_{h,i}\right] + (\tfrac{1}{2})\sum_i\sum_{j\neq i} \phi_{ti,tj}(r_{ij}), \qquad (11)$$

where ϕ now depends on the type of atom t_i and atom t_j. The host electron density is now given by,

$$\rho_{h,i} = \sum_{j\neq i} \rho^{*}{}_{tj}(r_{ij}), \qquad (12)$$

where the terms in the sum each depends on the type of neighbor atom j. Therefore, for a binary alloy with atom types A and B, the EAM energy requires definitions for $\phi_{AA}(r)$, $\phi_{BB}(r)$, $\phi_{AB}(r)$, $\rho_A(r)$, $\rho_B(r)$, $F_A(\rho)$ and $F_B(\rho)$.

1.2. The Many-Body Finnis and Sinclair (FS) Potentials: These potentials [86] were initially constructed to model the energetics of the transition metals. They avoid the problems associated with using pair potentials to model metals, e.g. the appearance of the Cauchy relation between the elastic constants $C_{12}=C_{44}$ which is not satisfied by cubic crystals. They also offer a better description of the surface relaxation in metals.

In the FS model, the total energy of an N-atom system is written as

$$\Phi_I^{FS} = (\tfrac{1}{2})\Sigma_{i \to N}\Sigma_{j \neq i}\phi_{rep}(r_{ij}) - c\Sigma_i\,(\rho_i)^{1/2}\,, \tag{13}$$

where

$$\rho_i = \Sigma_{j \neq i}\,\phi_{CE}(r_{ij})\,. \tag{14}$$

The function $\phi_{rep}(r_{ij})$ is a pairwise repulsive interaction between atoms i and j, separated by a distance r_{ij}, $\phi_{CE}(r_{ij})$ are two-body cohesive energy pair potentials and c is a positive constant. The second term in Eq. (13) represents the cohesive many-body contribution to the energy. The square root form of this term was motivated by an analogy with the second moment approximation to the Tight-Binding Model [101]. To see this, one starts with the tight-binding approach [102] in which the total electronic band energy, *i.e.* the total bonding energy, which is given as the sum of the energies of the occupied one-electron states, is expressed by

$$E_{tot} = 2 \int_{-\infty}^{E_f} E.n(E).dE\,, \tag{15}$$

where $n(E)$ is the electron density of states, E_f is the Fermi level energy and the factor 2 refers to spin degeneracy. E_{tot} is an attractive contribution to the configurational energy, which is dominated by the broadening of the partly filled valence shells of the atoms into bands when the solid is formed [103]. It is convenient to divide E_{tot} into contributions from individual atoms

$$E_{tot} = \Sigma_i E_i = 2 \int_{-\infty}^{E_f} \Sigma_i\ E.n_i(E)dE\,, \tag{16}$$

$$n_i(E) = \Sigma_\upsilon |< \Psi_\upsilon \mid i >|^2\,\delta\,(E - E_\upsilon)\,, \tag{17}$$

is the projected density of states on site i and $|\Psi_v>$ are the eigenfunctions of the one-electron Hamiltonian as has been discussed in [103]. To obtain $n_i(E)$ exactly, it is in principle necessary to know the positions of all atoms in the crystal. Furthermore, $n_i(E)$ is a very complicated functional of these positions. However, it is not necessary to calculate the detailed structure of $n_i(E)$. To obtain an approximate value of quantities such as E_i which involves integrals over $n_i(E)$, one needs only information about its width and gross features of its shape. This information is conveniently summarized in the moments of $n_i(E)$, defined by

$$\mu_n^{\ i} = \int\limits_{-\infty}^{\infty} E^n n_i(E) dE \qquad (18)$$

The important observation, which allows a simple description comparable to that of interatomic potentials, is that these moments are rigorously determined by the local environment. The exact relations are [103]

$$\mu^i_2 = \Sigma_j \, h_{ij}^{\ 2}$$
$$\mu^i_3 = \Sigma_{jk} \, h_{ij}h_{jk}h_{ki} \qquad (19)$$
$$\mu^i_4 = \Sigma_{jkl} \, h_{ij}h_{jk}h_{ki}h_{li},$$

where

$$h_{ij} = <\chi_i \,|\, H \,|\, \chi_j>, \qquad (20)$$

and χ_i is the localized orbital centered on atom i, and H is the one-electron Hamiltonian. Therefore, if an approximate expression for the E_i in terms of the first few μ^i_n is available, the electronic band energy can be calculated with essentially the same machinery used to evaluate interatomic potentials. Now, the exact evaluation of E_i requires the values of all the moments on site i. However, a great deal of information can be gained from a description based only on the second moment, μ^i_2. This moment provides a measure of the squared valence-band width, and thus sets a basic energy scale for the problem. Therefore, a description using only μ^i_2 assumes that the effects of the structure of $n_i(E)$ can be safely ignored, since the higher moments describe the band shape. Since, E_i has units of energy and $\mu^i{}_2$ has units of (energy)^2 , therefore one can write

$$E_i = E_i(\mu^i{}_2) = - A \sqrt{(\mu^i{}_2)} = - A \sqrt{(\Sigma_j h_{ij}{}^2)}, \qquad (21)$$

where A is a positive constant that depends on the chosen density of states shape and the fractional electron occupation [96].

The functions $\phi_{CE}(r_{ij})$ in Eq. (14) can be interpreted as the sum of squares of hopping (overlap) integrals. The function ρ_i can be interpreted as the local electronic charge density [76] constructed by a rigid superposition of the atomic charge densities $\phi_{CE}(r_{ij})$. In this interpretation, the energy of an atom at the site i is assumed to be identical to its energy within a uniform electron gas of that density. Alternatively, ρ_i, can be interpreted [86] as a measure of the local density of atomic sites, in which case Eq. (13) can be considered as a sum consisting of a part that is a function of the local volume, represented by the second term, and a pairwise interaction part, represented by the first term. The FS potentials, Eq. (13), are similar in form to the EAM potentials in Eq. (8). However, their interpretations are quite different. The FS potentials, as has been shown above, were derived on the basis of the Tight-Binding Model and this is the reason why their many-body parts, which correspond to the $F_i [\rho_{h,i}]$ functionals in the EAM potentials, are in the form of square root terms. Furthermore, the FS potentials are less convenient than the EAM potentials for a conversion from the pure metals to their alloys. Notwithstanding this difficulty, FS potentials have been constructed for several alloy systems, such as the alloys of the noble metals (Au, Ag, Cu) [87].

1.3. The Many-Body Sutton and Chen (SC) Long-Range Potentials:
The SC potentials [88] describe the energetics of ten FCC elemental metals. They are of the FS type and therefore similar in form to the EAM potentials. They were specifically designed for use in computer simulations of nanostructures involving a large number of atoms.

In the SC potentials, the total energy, written in analogy with Eq. (13), is given by

$$\Phi_I^{SC} = \varepsilon \left[(\tfrac{1}{2}) \sum_i \sum_{j \neq i} \phi_{rep}(r_{ij}) - c \sum_i (\rho_i)^{1/2} \right], \qquad (22)$$

where

$$\phi_{rep}(r_{ij}) = (a/r_{ij})^n, \qquad (23)$$

and

$$\rho_i = \sum_{j \neq i} (a/r_{ij})^m, \qquad (24)$$

where ε is a parameter with the dimensions of energy, a is a parameter with the dimensions of length and is normally taken to be the equilibrium lattice constant, m and n are positive integers with $n > m$. The power-law form of the potential terms was adopted so as to construct a unified model that can combine the short-range interactions, afforded by the N-body second term in Eq. (22) and useful for the description of surface relaxation phenomena, with a van der Waals tail that gives a better description of the interactions at the long range. For a particular FCC elemental metal, the potential in Eq. (22) is completely specified by the values of m and n, since the equilibrium lattice condition fixes the value of c. The values of the potential parameters, computed for a cut-off radius of 10 lattice constants, are listed in Table 3.

Table 3. Parameters of the Sutton-Chen potentials.

Element	m	n	$\varepsilon\,(eV)$	c
Ni	6	9	1.5707×10^{-2}	39.432
Cu	6	9	1.2382×10^{-2}	39.432
Rh	6	12	4.9371×10^{-3}	144.41
Pd	7	12	4.1790×10^{-3}	108.27
Ag	6	12	2.5415×10^{-3}	144.41
Ir	6	14	2.4489×10^{-3}	334.94
Pt	8	10	1.9833×10^{-2}	34.408
Au	8	10	1.2793×10^{-2}	34.408
Pb	7	10	5.5765×10^{-3}	45.778
Al	6	7	3.3147×10^{-2}	16.399

These parameters were obtained by fitting the experimental cohesive energies and lattice parameters exactly. The indices m and n were restricted to integer values, such that the product $m.n$ was the nearest integer to $18.\Omega^f.B^f/E^f$, Eq. (9) in [88], where Ω^f is the FCC atomic volume, B^f is the computed bulk modulus, and E^f is the fitted cohesive energy.

The SC potentials have been applied to the computation of the elastic constants, bulk moduli and cohesive energies of the FCC metals, and the prediction of the relative stabilities of the FCC, BCC and HCP structures [88]. The results show reasonable agreement with the experimental values. These potentials have also been used in modeling the structural properties of metallic clusters in the size range of 13 to 309 atoms [94].

1.4. The Many-Body Murrell–Mottram (MM) Potential: The Murrell-Mottram potentials is an example of cluster-type potentials, and consists of sums of effective two- and three body interactions [75,105,106]

$$\Phi_{tot} = \Sigma_i \Sigma_{j>i} \Phi_{ij}^{(2)} + \Sigma_i \Sigma_{j>i} \Sigma_{k>j} \Phi_{ijk}^{(3)}. \tag{25}$$

The pair interaction term is modeled by a Rydberg function which has been used for simple diatomic potentials. In the units of reduced energy and distance, it takes the form

$$\Phi_{ij}^{(2)}/D = -(1 + a_2 \rho_{ij})exp(-a_2 \rho_{ij}), \tag{26}$$

where

$$\rho_{ij} = (r_{ij} - r_e)/r_e. \tag{27}$$

D is the depth of the potential minimum, corresponding to the diatomic dissociation energy at $\rho_{ij}=0$, i.e. for $r_{ij}=r_e$, with r_e the diatomic equilibrium distance. D and r_e are fitted to the experimental cohesive energy and lattice parameter data, respectively. The only parameter involved in the optimization of the potential is a_2, which is related to the curvature (force constant) of the potential at its minimum [75,105,106]. The three-body term must be symmetric with respect to the permutation of the three atoms indices, i, j and k. The most convenient way to achieve this is to create functional forms which are combinations of interatomic coordinates, Q_1, Q_2 and Q_3 which are irreducible representations of the S3 permutation group [107]. If one constructs a given triangle with atoms (i, j, k), then the coordinates Q_i will be given by

$$\begin{bmatrix} Q_1 \\ Q_2 \\ Q_3 \end{bmatrix} = \begin{vmatrix} \sqrt{1/3} & \sqrt{1/3} & \sqrt{1/3} \\ 0 & \sqrt{1/2} & -\sqrt{1/2} \\ \sqrt{2/3} & -\sqrt{1/6} & -\sqrt{1/6} \end{vmatrix} \begin{bmatrix} \rho_{ij} \\ \rho_{jk} \\ \rho_{ki} \end{bmatrix}, \qquad (28)$$

with

$$\rho_{\alpha\beta} = (r_{\alpha\beta} - r_e)/r_e, \qquad (29)$$

where $r_{\alpha\beta}$ represents one of the three triangle edges (r_{ij}, r_{jk}, r_{ki}). These interatomic coordinates have specific geometrical meanings. Q_1 represents the perimeter of the triangle in reduced unit; Q_2 and Q_3 measure the distortions from an equilateral geometry [75]. All polynomial forms, which are totally symmetric in $\rho_{\alpha\beta}$ can be expressed as sums of products of the so-called integrity basis [75], defined as:

$$Q_1, \ Q_2^2 + Q_3^2, \ Q_3^3 - 3Q_3Q_2^2. \qquad (30)$$

A further condition that must be imposed on the three-body term is that it must go to zero if any one of the three atoms goes to infinity. The following general family of functions can be chosen for the three-body term to conform to the functional form adopted for the two-body term:

$$\Phi_{ijk}^{(3)}/D = P(Q_1,Q_2,Q_3).F(a_3,Q_1), \qquad (31)$$

where $P(Q_1,Q_2,Q_3)$ is a polynomial in the Q coordinates and F is a damping function, containing a single parameter, a_3, which determines the range of the three-body potential. Three different kinds of damping functions can be adopted:

$$F(a_3,Q_1) = exp(- a_3Q_1) \qquad \text{exponential,}$$
$$F(a_3,Q_1) = (\tfrac{1}{2})[1 - tanh(a_3Q_1/2)] \qquad \text{tanh,} \qquad (32)$$
$$F(a_3,Q_1) = sech(a_3Q_1) \qquad \text{sech.}$$

The use of the exponential damping function can lead to a problem, namely, for large negative Q_1 values (i.e. for triangles for which $r_{ij}+r_{jk}+r_{ki}<<3r_e$), the function F may be large so that the three-body contribution swamps the total two-body contribution. This may lead to the collapse of the lattice. To overcome this problem, it may be necessary in some cases, to add a hard wall function to the repulsive part of the two-body term.

The polynomial, P, is normally taken to be

$$P(Q_1,Q_2,Q_3) = c_o + c_1Q_1 + c_2Q_1^2 + c_3(Q_2^2+Q_3^2) + c_4Q_1^3$$
$$+ c_5Q_1(Q_2^3+Q_3^2) + c_6(Q_3^3-3Q_3Q_2^2). \qquad (33)$$

This implies that there are seven parameters to be determined. For systems where simultaneous fitting is made to data for two different solid phases the following fourth-order terms can be added

$$c_7Q_1^4 + c_8Q_1^2.(Q_2^2+Q_3^2) + c_9(Q_2^2+Q_3^2)^2 + c_{10}Q_1(Q_3^3-3Q_3Q_2^2). \qquad (34)$$

The potential parameters for a set of elements are given in Table 4.

Table 4. Parameters of the Murrell-Mottram potentials.

Metal	A_2	a_3	$D(eV)$	$r_e\phi$ [nm]	c_o	c_1	C_2
Al	7.0	8.0	0.9073	0.27568	0.2525	- 0.4671	4.4903
Cu	7.0	9.0	0.888	0.2448	0.202	-0.111	4.990
Ag	7.0	9.0	0.722	0.2799	0.204	-0.258	6.027
Sn	6.25	3.55	1.0	0.2805	1.579	-0.872	-4.980
Pb	8.0	6.0	0.59273	0.332011	0.18522	0.87185	1.27047

Table 4. continued.

Metal	c_3	c_4	c_5	c_6	c_7	c_8	c_9	c_{10}
Al	-1.1717	1.6498	-5.3579	1.6327	0.0	0.0	0.0	0.0
Cu	-1.369	0.469	-2.630	1.202	0.0	0.0	0.0	0.0
Ag	-1.262	-0.442	-5.127	2.341	0.0	0.0	0.0	0.0
Sn	-13.145	-4.781	35.015	-1.505	2.949	-15.065	10.572	12.830
Pb	-3.44145	-3.884	155.27033	2.85596	0.0	0.0	0.0	0.0

1.5. The Many-Body Rafii–Tabar and Sutton (RTS) Long-Range Alloy Potentials: In developing these potentials these authors [69,89] considered the case of many-body interatomic potentials that describe the energetics of metallic alloys, and in particular the FCC metallic alloys. The interatomic potential that models the energetics and dynamics of a binary, A-B, alloy is normally constructed from the potentials that separately describe the A-A and the B-B interactions, where A and B are the elemental metals. To proceed with this scheme, a combining rule is normally proposed. Such a rule would allow for the computation of the A-B interaction parameters from those of the A-A and B-B parameters. The combining rule reflects the different averaging procedures that can be adopted, such as the arithmetic or the geometric averaging. The criterion for choosing any one particular combining rule is the closeness of the results obtained, when computing with the proposed A-B potential with that rule, and the corresponding experimental values.

The Rafii-Tabar and Sutton potentials [69,89] are the generalization of the SC potentials and model the energetics of the metallic FCC random binary alloys. They have the advantage that all the parameters for the alloys are obtained from those for the elemental metals without the introduction of any new parameters. The basic form of the potential is given by

$$
\begin{aligned}
\Phi^{RTS} = {}& (\tfrac{1}{2})\Sigma_i\Sigma_{j\neq i}\ \hat{p}_i\ \hat{p}_j\ V^{AA}(r_{ij}) + (1-\hat{p}_i)(1-\hat{p}_j)V^{BB}(r_{ij}) \\
&+ [\ \hat{p}_i(1-\hat{p}_j) + \hat{p}_j(1-\hat{p}_i)]\ V^{AB}(r_{ij}) \\
&- d^{AA}\Sigma_i\ \hat{p}_i\ [\Sigma_{j\neq i}\ \hat{p}_j\ \Phi^{AA}(r_{ij}) + (1-\hat{p}_j)\Phi^{AB}(r_{ij})]^{1/2} \\
&- d^{BB}\Sigma_i(1-\hat{p}_i)\ [\Sigma_{j\neq i}(1-\hat{p}_j)\Phi^{BB}(r_{ij}) + \hat{p}_j\ \Phi^{AB}(r_{ij})]^{1/2}. \qquad (35)
\end{aligned}
$$

The operator $\hat{p}_i$ is the site occupancy operator and is defined as

$$\hat{p}_i = 1 \qquad \text{if site i is occupied by an A atom}$$
$$\hat{p}_i = 0 \qquad \text{if site i is occupied by a B atom} \qquad (36)$$

The functions $V^{\alpha\beta}$ and $\Phi^{\alpha\beta}$ are defined as

$$V^{\alpha\beta}(r) = \varepsilon^{\alpha\beta} \; [a^{\alpha\beta}/r]^{n\alpha\beta},$$

$$\Phi^{\alpha\beta}(r) = [a^{\alpha\beta}/r]^{m\alpha\beta} \qquad (37)$$

where ε and β are both A and B. The parameters ε^{AA}, c^{AA}, a^{AA}, m^{AA} and n^{AA} are for the pure element A, and ε^{BB}, c^{BB}, a^{BB}, m^{BB} and n^{BB} are for the pure element B, given in Table 3.

$$d^{AA} = \varepsilon^{AA} \, c^{AA},$$

$$d^{BB} = \varepsilon^{BB} \, c^{BB}. \qquad (38)$$

The mixed, or alloy, states are obtained from the pure states by assuming the combining rules:

$$V^{AB} = (V^{AA} V^{BB})^{\frac{1}{2}}, \qquad (39)$$

$$\Phi^{AB} = (\Phi^{AA} \Phi^{BB})^{\frac{1}{2}}. \qquad (40)$$

These combining rules, based on purely empirical grounds, give the alloy parameters as

$$m^{AB} = 1/2 \, (m^{AA} + m^{BB}),$$
$$n^{AB} = 1/2 \, (n^{AA} + n^{BB}),$$
$$a^{AB} = (a^{AA} a^{BB})^{\frac{1}{2}}, \qquad (41)$$
$$\varepsilon^{AB} = (\varepsilon^{AA} \varepsilon^{BB})^{\frac{1}{2}}.$$

These potentials were used to compute the elastic constants and heat of formation of a set of FCC metallic alloys [89], as well as to model the formation of ultra thin Pd films on Cu(100) surface [90]. They form the basis of a large class of MD simulations [69,70].

1.6. Angular-Dependent Potentials: Transition metals form three rather long rows in the Periodic Table, beginning with Ti, Zr and Hf and terminating with Ni, Pd and Pt. These rows correspond to the filling of 3d, 4d and 5d orbital shells, respectively. Consequently, the d-band interactions play an important role in the energetics of these metals [108], giving rise to angular-dependent forces that contribute significantly to the structural and vibrational characteristics of these elements. Pseudopotential models are commonly used to represent the intermolecular interaction in such metals [109,110]. Recently, an *ab initio* generalized pseudopotential theory [111] was employed to construct an analytic angular-dependent potential for the description of the element Mo [112], a BCC transition metal. According to this prescription, the total cohesive energy is expressed as

$$\Phi_I^{MO} = \Phi_{vol}(\Omega) + (\tfrac{1}{2N}) \Sigma_i \Sigma_{j \neq i} \ \Phi_2(_{ij})$$
$$+ \ (^1/_{6N}) \ \Sigma_i \Sigma_{j \neq i} \Sigma_{k \neq i,j} \ \Phi_3(_{ijk})$$
$$+ \ (^1/_{24N}) \ \Sigma_i \Sigma_{j \neq i} \Sigma_{k \neq i,j} \ \Sigma_{l \neq i,j,k} \ \Phi_4(_{ijkl}) \qquad (42)$$

where Ω is the atomic volume, N is number of ions, Φ_3 and Φ_4 are, respectively, the angular-dependent three- and four-ion potentials and Φ_{vol} includes all one-ion intraatomic contributions to the cohesive energy. The interatomic potentials, $\Phi_2(_{ij})$, $\Phi_3(_{ijk})$ and $\Phi_4(_{ijkl})$ denote

$$\Phi_2(_{ij}) \equiv \Phi_2(r_{ij}; \ \Omega) \, ,$$
$$\Phi_3(_{ijk}) \equiv \Phi_3(r_{ij}, \ r_{jk}, \ r_{kl}; \ \Omega) \, , \qquad (43)$$
$$\Phi_4(_{ijkl}) \equiv \Phi_4(r_{ij}, \ r_{jk}, \ r_{kl}, \ r_{li}, \ r_{ki}, \ r_{li}; \ \Omega) \, ,$$

where r_{ij}, for example, is the ion-ion separation distance between ions i and j. These potentials are expressible in terms of weak pseudopotential and d-state tight-binding and hybridization matrix elements that couple different sites. Analytic expressions for these functions are provided [111,112] in terms of distances and angles subtended by these distances.

The potential expressed by Eq. (42) was employed to compute the values of a set of physical properties of Mo including the elastic constants, the phonon frequencies and the vacancy formation energy [112]. These results clearly show that the inclusion of the angular-dependent potentials greatly improves the computed values of these properties as compared with the results obtained exclusively from an effective two-body interaction potential, V_2^{eff}. Furthermore, the potential was employed in an MD simulation of the melting transition of the Mo, details of which can be found in [112].

2. Interatomic Potentials for Covalently-Bonding Systems

2.1. The Tersoff Many-Body C–C, Si–Si and C–Si Potentials:
The construction of Tersoff many-body potentials are based on the formalism of analytic bond-order potential, initially suggested by Abell [113]. According to Abell's prescription, the binding energy of an atomic many-body system can be computed in terms of pairwise nearest-neighbor interactions that are, however, modified by the local atomic environment. Tersoff employed this prescription to obtain the binding energy in Si [114-116], C [117], Si-C [116,118], Ge and Si-Ge [118] solid-state structures.

In the Tersoff's model, the total binding energy is expressed as

$$\Phi_I^{TR} = \Sigma_i E_i = (\frac{1}{2})\Sigma_i\Sigma_{j\neq i}\phi(r_{ij}), \qquad (44)$$

where E_i is the energy of site i and $\phi(r_{ij})$ is the interaction energy between atoms i and j, given by

$$\phi(r_{ij}) = f_c(r_{ij}) [\phi_{rep}(r_{ij}) + b_{ij}\phi_{att}(r_{ij})] . \qquad (45)$$

The function $V_{rep}(r_{ij})$ represents the repulsive pairwise potential, such as the core-core interactions, and the function $V_{att}(r_{ij})$ represents the attractive bonding due to the valence electrons. The many-body feature of the potential is represented by the term b_{ij} which acts as the bond-order term and which depends on the local atomic environment in which a particular bond is located. The analytic forms of these potentials are given by

$$\phi_{rep}(r_{ij}) = A_{ij} \, exp(- \lambda_{ij} \, r_{ij}),$$
$$\phi_{att}(r_{ij}) = - B_{ij} \, exp(- \mu_{ij} \, r_{ij}),$$

$f_c(r_{ij}) = 1$,   for $r_{ij} < R_{ij}^{(1)}$,

$f_c(r_{ij}) = (\frac{1}{2})+(\frac{1}{2})cos[\pi(r_{ij}-R_{ij}^{(1)})/(R_{ij}^{(2)}-R_{ij}^{(1)})]$,   for $R_{ij}^{(1)} < r_{ij} < R_{ij}^{(2)}$,

$f_c(r_{ij}) = 0$,   for $r_{ij} > R_{ij}^{(2)}$

$$b_{ij} = \chi_{ij}[1 + (\beta_i \zeta_{ij})^{ni}]^{-0.5ni},$$
$$\zeta_{ij} = \Sigma_{k\neq i,j} f_c(r_{ik})\omega_{ik}\, g(\theta_{ijk}),$$
$$g(\theta_{ijk}) = 1 + c_i^2/d_i^2 - c_i^2/[d_i^2 + (h_i - cos\theta_{ijk})^2],$$
$$\lambda_{ij} = (\lambda_i + \lambda_j)/2,$$
$$\mu_{ij} = (\mu_i + \mu_j)/2,$$
$$\omega_{ik} = exp[\mu_{ik}(r_{ij} - r_{ik})]^3,$$
$$A_{ij} = \sqrt{A_iA_j},$$
$$B_{ij} = \sqrt{B_iB_j},$$
$$R_{ij}^{(1)} = \sqrt{R_i^{(1)}R_j^{(1)}},$$
$$R_{ij}^{(2)} = \sqrt{R_i^{(2)}R_j^{(2)}}, \tag{46}$$

Numerical values of the parameters of Tersoff potentials for C and Si are listed in Table 5 where the labels i, j and k refer to the atoms in the ijk bonds, r_{ij} and r_{ik} refer to the lengths of the ij and ik bonds whose angle is θ_{ijk}. Singly subscripted parameters, such as λ_i and n_i, depend only on one type of atom, e.g. C or Si. The parameters for the C-C, Si-Si and Si-C potentials are listed in Table 5. For the C, the parameters were obtained by fitting the cohesive energies of carbon polytypes, along with the lattice constant and bulk modulus of diamond. For the Si, the parameters were obtained by fitting to a database consisting of cohesive energies of real and hypothetical bulk polytypes of Si, along with the bulk modulus and bond length in the diamond structure. Furthermore, these potential parameters were required to reproduce all three elastic constants of Si to within 20%.

Table 5. Parameters of the Tersoff potentials for C and Si.

Parameter	C	Si
A(ev)	1.3936×10^3	1.8308×10^3
B(ev)	3.467×10^2	4.7118×10^2
λ (ϕ [nm])$^{-1}$	34.879	24.799
μ (ϕ [nm])$^{-1}$	22.119	17.322
β	1.5724×10^{-7}	1.1000×10^{-6}

Table 5. continued.

Parameter	C	Si
η	7.2751×10^{-1}	7.8734×10^{-1}
c	3.8049×10^4	1.0039×10^5
d	4.384	16.217
h	-0.57058	-0.59825
$R^{(1)}$ (ϕ [nm])	0.18	0.27
$R^{(2)}$ (ϕ [nm])	0.21	0.30
χ	1	1
$\chi_{C\text{-}Si}$	0.9776	

2.2. The Brenner–Tersoff-Type First Generation Hydrocarbon Potentials:

The Tersoff potentials correctly model the dynamics of a variety of solid-state structures, such as the surface reconstruction in Si [114,115] or the formation of interstitial defects in carbon [117]. However, while these potentials can give a realistic description of the C-C single, double and triple bond lengths and energies in hydrocarbons, solid graphite and diamond, they lead to non-physical results for bonding situations intermediate between the single and double bonds.   One example is the bonding in the Kekule construction for the graphite where, due to bond conjugation, each bond is considered to be approximately one-third double-bond and two-thirds single-bond in character. To correct for this, and similar problems in hydrocarbons, as well as to correct for the non-physical overbinding of radicals, Brenner [119] developed a Tersoff-type potential for hydrocarbons that can model the bonding in a variety of small hydrocarbon molecules as well as in diamond and graphite. In this potential, Eqs. (44) and (45) are written as

$$\Phi_I^{Br} = (\tfrac{1}{2})\sum_i \sum_{i \neq j} \phi(r_{ij}) \qquad (47)$$

and

$$\phi(r_{ij}) = f_c(r_{ij}) \left[\phi_{rep}(r_{ij}) + \bar{b}_{ij} \, \phi_{att}(r_{ij}) \right], \qquad (48)$$

where

$$\phi_{rep}(r_{ij}) = D_{ij}/(S_{ij}-1).exp[-\sqrt(2S_{ij}).\beta_{ij} \, (r_{ij} - R_{ij}^e)] ,$$

$$\phi_{att}(r_{ij}) = -D_{ij}S_{ij}/(S_{ij}-1).exp[-\sqrt{(2S_{ij})}.\beta_{ij}(r_{ij}-R_{ij}^{e})]\,,$$

$$\overline{b}_{ij} = (b_{ij}+b_{ji})/2 + F_{ij}(N_i^{(t)},N_j^{(t)},N_{ij}^{Conj})\,,$$

$$b_{ij} = [1 + G_{ij} + H_{ij}(N_i^{(H)},N_i^{(C)})]^{-\delta i}\,,$$

$$G_{ij} = \Sigma_{k\neq i,j}f_c(r_{ik})\,G_i(\theta_{ijk}).exp[\alpha_{ijk}\{(r_{ij}-R_{ij}^{(e)}) - (r_{ik}-R_{ik}^{(e)})\}]\,,$$

$$G_c(\theta) = a_o[1 + c_o^2/d_o^2 - c_o^2/[d_o^2 + (1+cos\theta)^2]\,. \tag{49}$$

The quantities $N_i^{(C)}$ and $N_i^{(H)}$ represent the number of C and H atoms bonded to atom i. $N_i^{(t)}=(N_i^{(C)}+N_i^{(H)})$ is the total number of neighbors of atom i and its values, for neighbors of the two carbon atoms involved in a bond, can be used to determine if the bond is part of a conjugated system. For example, if $N_i^{(t)}<4$, then the carbon atom forms a conjugated bond with its carbon neighbors. N_{ij}^{conj} depends on whether an ij carbon bond is part of a conjugated system. These quantities are given by

$$N_i^{(H)} = \overset{\text{hydrogen atoms}}{\Sigma_{l\neq i,j}f_c(r_{il})}\,,$$

$$N_i^{(C)} = \overset{\text{carbon atoms}}{\Sigma_{k\neq i,j}f_c(r_{ik})}\,,$$

$$N_{ij}^{conj} = 1 + \overset{\text{carbon atoms}}{\Sigma_{k\neq i,j}f_c(r_{ik})F(x_{ik})} + \overset{\text{carbon atoms}}{\Sigma_{l\neq i,j}f_c(r_{jl})F(x_{jl})}\,,$$

$$F(x_{ik}) = 1\,, \qquad\qquad \text{for} \quad x_{ik}\leq 2$$
$$F(x_{ik}) = \tfrac{1}{2}+ (\tfrac{1}{2})cos[\pi(x_{ik}-2)]\,, \quad \text{for} \quad 2<x_{ik}<3$$
$$F(x_{ik}) = 0\,, \qquad\qquad \text{for} \quad x_{ik}\geq 3$$

$$x_{ik} = N_k^{(t)} - f_c(r_{ik})\,. \tag{50}$$

The expression for N_{ij}^{conj} yields a continuous value as the bonds break and form, and as the second-neighbor coordinations change. For $N_{ij}^{conj}=1$ the bond between a pair of carbon atoms i and j is not part of a conjugated system, whereas for $N^{conj}\geq 2$ the bond is part of a conjugated system.

The functions H_{ij} and F_{ij} are parameterized by two- and three-dimensional cubic splines, respectively, and the potential parameters in Eqs. (47) to (50) were determined by first fitting to systems composed of carbon and hydrogen atoms only, and then the parameters were chosen for the mixed hydrocarbon systems. Two sets of parameters, consisting of 63 and 64 entries, are listed in [119]. These parameters were obtained by fitting a variety of hydrocarbon data sets, such as the binding energies and lattice constants of graphite, diamond, simple cubic and FCC structures, and the vacancy formation energies. The complete fitting sets are given in Tables 3-5 in [119].

2.3. The Brenner–Tersoff-Type Second Generation Hydrocarbon Potentials:
The potential function, expressed by Eqs. (47)-(50) and referred to as the first generation hydrocarbon potential, was recently further refined [73, 120] by including improved analytic functions for the intramolecular interactions, and by an extended fitting database, resulting in a significantly better description of bond lengths, energies and force constants for hydrocarbon molecules, as well as elastic properties, interstitial defect energies, and surface energies for diamond. In this improved version, the terms in Eq. (48) are redefined as

$$\phi_{rep}(r_{ij}) = f_c(r_{ij}).[1 + Q_{ij}/r_{ij}] \, A_{ij}.exp(\alpha_{ij} \, r_{ij}) \, ,$$

$$\phi_{att}(r_{ij}) = - f_c(r_{ij}) \, \Sigma_{(n=1,3)} \, B_{ijn}.exp(\beta_{ijn} \, r_{ij}) \, ,$$

$$\bar{b}_{ij} = (p_{ij}^{\,\sigma\pi} + p_{ji}^{\,\sigma\pi})/2 + p_{ij}^{\,\pi} \, ,$$

$$p_{ij}^{\,\pi} = \pi_{ij}^{\,rc} + \pi_{ij}^{\,dh} \, ,$$

$$p_{ij}^{\,\sigma\pi} = [1 + G_{ij} + P_{ij}(N_i^{(H)}, N_i^{(C)})]^{\,-\frac{1}{2}} \, ,$$

$$G_{ij} = \Sigma_{k \neq i,j} f_c(r_{ik}) G_i \, [cos(\theta_{jik})].exp[\lambda_{ijk}(r_{ij} - r_{ik}) \, ,$$

$$\pi_{ij}^{\,rc} = F_{ij} \, (N_i^{(t)}, N_j^{(t)}, N_{ij}^{\,conj}) \, ,$$

$$N_{ij}^{\,conj} = 1 + [\, \Sigma_{k \neq i,j} f_c(r_{ik}) F(x_{ik})]^{\,2} \overset{carbon \; atoms}{} + [\, \Sigma_{l \neq i,j} f_c(r_{jl}) F(x_{jl})]^{\,2} \overset{carbon \; atoms}{} ,$$

$$\pi_{ij}^{dh} = T_{ij}\ (N_i^{(t)}, N_j^{(t)}, N_{ij}^{conj}).[\textstyle\sum_{k\ne i,j}\sum_{l\ne i,j}\ (1 - \cos^2\omega_{ijkl})f_c(r_{ik}).f_c(r_{jl})],$$

$$\cos\omega_{ijkl} = e_{ijk}.e_{ijl} .\tag{51}$$

Q_{ij} is the screened Coulomb potential, which goes to infinity as the interatomic distances approach zero. The term π_{ij}^{rc} represents the influence of radical energetics and π-bond conjugation on the bond energies, and its value depends on whether a bond between atoms i and j has a radical character and is part of a conjugated system. The value of π_{ij}^{dh} depends on the dihedral angle for the C-C double bonds. P_{ij} represents a bicubic spline, F_{ij} and T_{ij} are tricubic spline functions. In the dihedral term, π_{ij}^{dh}, the functions e_{jik} and e_{ijl} are unit vectors in the direction of the cross products $R_{ji} \times R_{ik}$ and $R_{ij} \times R_{jl}$, respectively, where the R's are the interatomic vectors. The function $G_c[\cos(\theta_{jik})]$ modulates the contribution that each nearest-neighbor makes to $\bar{b}_{ij}$. This function was determined in the following way. It was computed for the selected values of $\theta=109.47^o$ and $\theta=120^o$, corresponding to the bond angles in diamond and graphitic sheets, and for $\theta= 90^o$ and $\theta= 180^o$, corresponding to the bond angles among the nearest neighbors in a simple cube lattice. The FCC lattice contains angles of 60°, 90°, 120° and 180°. A value of $G_c[\cos(\theta= 60^o)]$ was also computed from the above values. To complete an analytic function for the $G_c[\cos(\theta)]$, sixth order polynomial splines in $\cos(\theta)$ were used to obtain its values for θ between 109.47^o and 120°. For θ between 0° and 109°, for a carbon atom i, the angular function

$$g_c = G_c[\cos(\theta)] + Q(N_i^{(t)}).[\gamma_c \cos(\theta) - G_c[\cos(\theta)]] ,\tag{52}$$

is employed, where $\gamma_c \cos(\theta)$ is a second spline function, determined for angles less than 109.47^o. The function $Q(N_i^{(t)})$ is defined by

$$Q(N_i^{(t)}) = 1, \qquad\qquad\qquad\qquad \text{for}\quad N_i^{(t)} \le 3.2,$$

$$Q(N_i^{(t)}) = \tfrac{1}{2}+(\tfrac{1}{2})\cos[\pi(N_i^{(t)}-3.2)/(3.7-3.2)] , \quad \text{for}\quad 3.2 < N_i^{(t)} < 3.7 ,\tag{53}$$

$$Q(N_i^{(t)}) = 0, \qquad\qquad\qquad\qquad \text{for}\quad N_i^{(t)} \ge 3.7$$

The large database of the numerical data on parameters and spline functions were obtained by fitting the elastic constants, vacancy formation energies and the formation energies for interstitial defects for diamond.

3. Interatomic Potential for C–C Non-Covalent Systems

The non-covalent interactions between carbon atoms are required in many of the simulation studies in computational nanoscience and nanotechnology. These can be modeled according to various types of potential [10]. The Lennard-Jones and Kihara potentials can be employed to describe the van der Waals intermolecular interactions between carbon clusters, such as C_{60} molecules, and between the basal planes in a graphite lattice. Other useful potentials are the exp-6 potential [121], which also describes the C_{60}-C_{60} interactions, and the Ruoff-Hickman potential [122], which models the C_{60}-graphite interactions.

3.1. The Lennard–Jones and Kihara Potentials: The total interaction potential between the carbon atoms in two C_{60} molecules, or between those in two graphite basal planes, could be represented by the Lennard-Jones potential [123]

$$\Phi_I^{LJ}(r_{ij}^{IJ}) = 4\varepsilon\Sigma_i\Sigma_{j>i} \left[(\sigma/r_{ij}^{IJ})^{12} - (\sigma/r_{ij}^{IJ})^6\right], \qquad (54)$$

where I and J denote the two molecules (planes), r_{ij} is the distance between the atom i in molecule (plane) I and atom j in molecule (plane) J. The parameters of this potential, $(\varepsilon=0.24127 \times 10^{-2}\ ev$, $\sigma=0.34\ \phi$ $[nm])$, were taken from a study of graphite [124]. The Kihara potential is similar to the Lennard-Jones except for the fact that a third parameter d, is added to correspond to the hard-core diameter, i.e.

$$\Phi_I^{LJ}(r_{ij}^{IJ}) = 4\varepsilon\Sigma_i\Sigma_{j>i} \left[\{(\sigma-d)/(r_{ij}^{IJ}-d)\}^{12} - \left[\{(\sigma-d)/(r_{ij}^{IJ}-d)\}^6\right]\right]\ for\ r>d\ ,$$

$$\Phi_I^{LJ}(r_{ij}^{IJ}) = \infty \qquad\qquad\qquad for\ r \leq d \qquad (54\text{-}1)$$

3.2. The exp-6 Potential: This is another potential that describes the interaction between the carbon atoms in two C_{60} molecules

$$\Phi_I^{EXP6}(r_{ij}^{IJ}) = \Sigma_i\Sigma_{j>i} \, [A \, exp(- \alpha r_{ij}^{IJ})- B \, /(r_{ij}^{IJ})^6].$$ (55)

Two sets of values of the parameters are provided, and these are listed in Table 6. These parameters have been obtained from the gas phase data of a large number of organic compounds, without any adjustment.

Table 6. Parameters of the exp-6 potential for C.

	A(kcal/mol)	B [kcal/mol×$(\phi \, [nm])^6$]	α $(\phi \, [nm])^{-1}$
Set one	42000	3.58 x 10^8	35.8
Set two	83630	5.68 x 10^8	36.0

The measured value of the C_{60} solid lattice constant is a = 1.404 ϕ [nm] at T = 11° K. The calculated value using the set one was a = 1.301 ϕ [nm] and using the set two was a = 1.403ϕ [nm]. The experimentally estimated heat of sublimation is equal to - 45kcal/mol (extrapolated from the measured value of - 40.1 ± 1.3 kcal/mol at T = 707° K). The computed value using the set one was - 41.5 kcal/mol and using the set two was - 58.7 kcal/mol. We see that whereas the set two produces a lattice constant nearer the experimental value, the thermal properties are better described by using the set one.

3.3. The Ruoff–Hickman Potential: This potential, based on the model adopted by Girifalco [125], describes the interaction of a C_{60} molecule with a graphite substrate by approximating these two systems as continuum surfaces on which the carbon atoms are 'smeared out' with a uniform density. The sums over the pair interactions are then replaced by integrals that can be evaluated analytically. The C_{60} is modeled as a hollow sphere having a radius b = 0.355ϕ [nm], and the C-C pair interaction takes on a Lennard-Jones form

$$\phi_I(r_{ij}) = c_{12}r^{-12} - c_6r^{-6},$$ (56)

with $c_6 = 1.997 \times 10^{-5}$ $[ev.(\phi \, [nm])^6]$ and $c_{12} = 3.4812 \times 10^{-8}$ $[ev.(\phi \, [nm])^{12}]$ [125]. The interaction potential between the hollow C_{60} and a single carbon atom of a graphite substrate, located at a distance $z > b$ from the center of the sphere, is then evaluated as

$$\phi(z) = \phi_{12}(z) - \phi_6(z), \qquad (57)$$

where

$$\phi_n(z) = c_n/[2(n-2)].[N/(bz)].[1/(z-b)^{n-2} - 1/(z+b)^{n-2}], \qquad (58)$$

where N is the number of atoms on the sphere ($N = 60$ in this case) and $n = 12, 6$. The total interaction energy between the C_{60} and the graphite plane is then obtained by integrating $\phi(z)$ over all the atoms in the plane, giving

$$\phi_t(R) = E_{12}(R) - E_6(R), \qquad (59)$$

where

$$E_n(R) = \{c_n/[4(n-2)(n-3)]\}.(N^2/b^3).[1/(R-b)^{n-3} - 1/(R+b)^{n-3}], \qquad (60)$$

and R is the vertical distance of the center of the sphere from the plane.

4. Interatomic Potential for Metal–Carbon System

In modeling the growth of metallic films on semi-metallic substrates, such as graphite, a significant role is played by the interface metal-carbon potential since it controls the initial wetting of the substrate by the impinging atoms and also determines the subsequent diffusion and the final alignments of these atoms [10]. This potential has not been available and an approximate scheme is used, based on a combining rule, to derive its general analytic form [126]. To construct a mixed potential to describe the interaction of an FCC metallic atom (M) with C, a generalized Morse-like potential energy function is assumed,

$$\Phi_I^{MC}(r_{ij}) = \Sigma_i \Sigma_{j>i} E_{MC}[exp\{-N\alpha(r_{ij} - r_w)\} - N.exp\{-\alpha(r_{ij} - r_w)\}], \quad (61)$$

and to obtain its parameters, a known Morse potential function is employed,

$$\Phi_I^{CC}(r_{ij}) = \Sigma_i \Sigma_{j>i} E_C [exp\{- 2\alpha_1(r_{ij} - r_d)\} - 2exp\{- \alpha_1(r_{ij} - r_d)\}], \quad (62)$$

that describes the C-C interactions [127], and a generalized Morse-like potential function

$$\Phi_I^{MM}(r_{ij}) = \Sigma_i \Sigma_{j>i} E_M [exp\{- m\alpha_2(r_{ij} - r_o)\} - m.exp\{- \alpha_2(r_{ij} - r_o)\}], \quad (63)$$

that describes the M-M interactions [128]. Several combining rules were then tried. The rule giving the satisfactory simulation results led to

$$E_{MC} = \sqrt{E_C E_M},$$
$$r_w = \sqrt{r_d r_o},$$
$$\alpha = \sqrt{\alpha_1 \alpha_2}, \quad (64)$$
$$N = \sqrt{2m}.$$

Since a cut-off is normally applied to an interaction potential, the zero of this potential at a cut-off, r_c, was obtained according to the prescription in [127] leading to

$$\Phi_I^{MC}(r_{ij}) = \Sigma_i \Sigma_{j>i} E_{MC} [exp\{- N \alpha (r_{ij} - r_w)\} - Nexp\{- \alpha(r_{ij} - r_w)\}]$$
$$- E_{MC} [exp\{- N\alpha(r_c - r_w)\} - Nexp\{- \alpha(r_c - r_w)\}]$$
$$- E_{MC}N\alpha/\eta [1 - exp\eta(r_{ij} - r_c)]$$
$$\times [exp(-N\alpha(r_c - r_w)) - exp(-\alpha(r_c - r_w))], \quad (65)$$

where η is a constant whose value was chosen to be $\eta = 20$. This was a sufficiently large value so that the potential (84) was only modified near the cut-off distance. The parameters, pertinent to the case when the metal atoms were silver, i.e M=Ag, are listed in Table 7.

Table 7. Parameters of the Ag-C potential.

α_1	49.519 $(\phi\,[nm])^{-1}$
α_2	3.7152 $(\phi\,[nm])^{-1}$
E_C	3.1 ev
E_{Ag}	0.0284875 ev
M	6.00
r_o	0.444476 $\phi\,[nm]$
r_d	0.12419 $\phi\,[nm]$

The parameters for Eq. (62) were obtained by fitting the experimental cohesive energy and the inter-planar spacing, c /2 , of the graphite exactly, and the parameters for Eq. (63) were obtained by fitting the experimental values of the stress-free lattice parameter and elastic constants C_{11} and C_{12} of the metal.

5. Atomic-Site Stress Field

In many modeling studies involving the mechanical behavior of nanostructures, such as the simulation of the dynamics of crack propagation in an atomic lattice, it is necessary to compute a map of the stress distribution over the individual atomic sites in a system composed of N atoms [10].

The concept of atomic-level stress field was developed by Born and Huang [129] using the method of small homogeneous deformations. Applying small displacements to a pair of atoms i and j, with an initial separation of r_{ij}, it can be shown that [130] the Cartesian components of the stress tensor at the site i are given by

$$\sigma_{\alpha\beta}(i) = (1/2\Omega_i)\Sigma_{j>i}[\,\partial\Phi(r_{ij})/\partial r_{ij}\,].[r_{ij}^{\alpha}r_{ij}^{\beta}/r_{ij}]\,, \qquad (66)$$

where $\alpha, \beta = x, y, z$, $\Phi(r_{ij})$ is the two-body central potential, and Ω_i is the local atomic volume which can be identified with the volume of the Voronoi polyhedron associated with the atom i [131].

For the many-body potential energy given by Eq. (35), the stress tensor is given by

$$\sigma_{\alpha\beta}{}^{RTS}(i) = (\tfrac{1}{2}\Omega_i)[\Sigma_{j\neq i}[\partial V(r_{ij})/\partial r_{ij})]$$
$$- (\tfrac{1}{2})d^{AA}\,\hat{p}_i\,\Sigma_{j\neq i}\,(1/\sqrt{\rho_i^A} + 1/\sqrt{\rho_j^A}).[\partial\Phi^A\,(r_{ij})/\partial r_{ij}]$$
$$- (\tfrac{1}{2})d^{BB}(1 - \hat{p}_i\,)\,\Sigma_{j\neq i}\,(1/\sqrt{\rho_i^B}$$
$$+ 1/\sqrt{\rho_j^B}).[\partial\Phi^B(r_{ij})/\partial r_{ij}]].(r_{ij}{}^\alpha r_{ij}{}^\beta)/r_{ij}\,, \qquad (67)$$

which for an elemental lattice with the two-body potentials given in [132] reduces to (see also [133])

$$\sigma_{\alpha\beta}{}^{RTS}(i) = (\varepsilon/a^2)(\tfrac{1}{2}\Omega_i)[\Sigma_{j\neq i}\,[- n(a/r_{ij}\,)^{n+2} + cm(1/\sqrt{\rho_i}$$
$$+ 1/\sqrt{\rho_j})(a/r_{ij}\,)^{m+2}]\,(r_{ij}{}^\alpha r_{ij}{}^\beta)\,, \qquad (68)$$

where only the contribution of the virial component to the stress field has been included and the contribution of the kinetic energy part (momentum flux) has been ignored since the interest is only in the low-temperature stress distributions. The volumes associated with individual atoms, Ω_i, can be obtained by computing numerically their corresponding Voronoi polyhedra according to the prescription given in [134].

Conclusions and Discussion

In this chapter we have presented a review of direct measurement of interparticle force-distance relationship from which intermolecular potential energy functions data can be generated along with the methodology in using quantum mechanical *ab initio* calculation for generation of interparticle potential energy data. We have also presented a set of state-of-the-art phenomenological interatomic and intermolecular potential energy functions that are widely used in computational modeling at the nanoscale when several hundred to several millions of particles are involved. There is still a great deal of work need to be done in order to develop a thorough database for interatomic and intermolecular potential energy functions to be sufficient for applications in nanoscience and nanotechnology. Since to control the matter atom by atom, molecule by molecule and/or at the macromolecular level, which is the aim of the nanotechnology, it is necessary to know the exact intermolecular forces between the particles under consideration. In the development of intermolecular force models applicable for the study of nanostructures which are at the confluence of the smallest of human-made devices and the largest molecules of living systems it is necessary to reexamine the existing techniques and come up with more appropriate intermolecular force models.

It is understood that formidable challenges remain in the fundamental understanding of various phenomena in nanoscale before the potential of nanotechnology becomes a reality. With the knowledge of better and more exact intermolecular interactions between atoms and molecules it will become possible to increase our fundamental understanding of nanostructures. This will allow development of more controllable processes in nanotechnology and optimization of production and design of more appropriate nanostructures and their interactions with other nanosystems.

Bibliography

[1]. J.C. Phillips, Covalent Bonding in Crystals and Molecules and Polymers (Chicago Lectures in Physics), Univ. of Chicago Press; (January 1, 1970).

[2]. J. Whitesell and M. Fox, Organic Chemistry Module: Chapter 19 - Noncovalent Interactions and Molecular Recognition, Jones & Bartlett Pub; 2nd edition (January 1, 1997).

[3]. G.A. Mansoori and J.M. Haile, "Molecular Study of Fluids: A Historical Survey" Avd. Chem. Series 204, ACS, Wash., D.C., 1983.

[4]. J.M. Haile and G.A. Mansoori (Editors) "*Molecular-Based Study of Fluids*", Avd. Chem. Series 204, ACS, Wash., D.C. (1983).

[5]. W. Gans and J.C.A. Boeyens (Editors) "*Intermolecular Interactions*" Plenum Pub Corp; From Book News, Inc., October (1998).

[6]. P.L. Huyskens and W.A.P. Luck, Zeegers-Huyskens, T. (Editor), "*Intermolecular Forces: An Introduction to Modern Methods and Results*" Springer Verlag, New York, N.Y., October, (1991).

[7]. A. Stone, *The Theory of Intermolecular Forces* (International Series of Monographs on Chemistry, No. 32, Oxford University Press, Oxford, (1997).

[8]. K. Terakura and H. Akai, "*Interatomic Potential and Structural Stability*", Proceedings of the 15th Taniguchi Symposium," Kashikojima, Japan, Springer Series, (1993).

[9]. M. Edalat, F. Pang, S.S. Lan and G.A. Mansoori, *International Journal of Thermophysics*, 1, 2, 177-184 (1980).

[10]. H. Rafii-Tabar and G.A. Mansoori, "*Interatomic Potential Models for Nanostructures*" Encyclopedia Nanoscience & Nanotechnology, American Scientific Publishers (to appear), (2004).

[11]. F. Kermanpour, G.A. Parsafar, and G.A. Mansoori, *International Journal of Thermophysics,Vol.25,No.1,January 2004*. "Investigation of the Temperature and Density Dependences of the Effective Pair Potential Parameters Using Variational Theory"

[12]. A.R. Massih and G.A. Mansoori, *Fluid Phase Equilibria* 10, 57-72 (1983).

[13]. B.M. Axilrod and E. Teller, 11, 711, 299 (1943); see also E.H. Benmekki and G.A. Mansoori, *Fluid Phase Equi.* 41, 43 (1988).

[14]. G.A. Mansoori, *Fluid-Phase Equil.* 4, 61 (1980).

[15]. F. Firouzi and Modarress, H. and Mansoori, G.A., *Europ. Polymer J.* 34, 1489 (1998).

[16]. E.Z. Hamad and Mansoori, G.A., *J. Phys. Chem.* 94, 3148 (1990).

[17]. M.J. Frisch, Trucks, G.W. , Schlegel, H.B., Scuseria, G.E., Robb, M.A., Cheeseman, J.R., Zakrzewski, V.G., Montgomery, Jr., J.A., Stratmann, R.E., Burant, J.C., Dapprich, S., Millam, J.M., Daniels, A.D., Kudin, K.N., Strain, M.C., Farkas, O., Tomasi, J., Barone, V., Cossi, M., Cammi, R., Mennucci, B., Pomelli, C., Adamo, C., Clifford, S., Ochterski, J., Petersson, Ayala, G.A., Q. Cui, P.Y., Morokuma, K., Salvador, P., Dannenberg, J.J., Malick, D.K., Rabuck, A.D., Raghavachari, K., Foresman, J.B., Cioslowski, J., Ortiz, J.V., Baboul, A.G., Stefanov, B.B., Liu, G., Liashenko, A., Piskorz, P., Komaromi, I., Gomperts, R., Martin, R.L., Fox, D.J., Keith, T., Al-Laham, M.A., Peng, C.Y., Nanayakkara, A., Challacombe, M., Gill, P.M.W., Johnson, B., Chen, W., Wong, M.W., Andres, J.L., Gonzalez, C., Head-Gordon, M., Replogle E.S. and Pople, J.A., Gaussian, Inc., Pittsburgh, PA, (2001).

[18]. MOLPRO, A package of *ab initio* programs written by Werner, H.-J. and Knowles, P.J. with contributions from Almlf, J., Amos, R.D., Deegan, M.J.O., Elbert, S.T., Hampel, C., Meyer, W., Peterson, K., Pitzer, R., Stone, A.J., Taylor, P.R., Lindh, R., Mura, M.E. and Thorsteinsson, T.

[19]. A.K. Sum, S.I. Sandler, R. Bukowski and K. Szalewicz, *Journal of Chemical Physics*, 116, 7627-7636 (2002).

[20]. K. Esfarjani and G.A. Mansoori, "Statistical Mechanical Modeling and its Application to Nanosystems", Handbook of Theoretical and Computational Nanoscience and Nanotechnology (to appear) 2004.

[21]. W.R Busing (Editor), *"Intermolecular Forces and Packing in Crystals"* (Transactions of the American Crystallographic Association Series: Volume 6), Polycrystal Book Service; Huntsville, AL, December (1970).

[22]. J.G. Kirkwood, *"Dielectrics Intermolecular Forces Optical Rotation"*, Gordon & Breach Science Pub; June (1965).

[23]. H. Margenau, *"Theory of intermolecular forces"* Pergamon Press; 2d ed., (1971).

[24]. J. Israelachvili, *"Intermolecular and Surface Forces: With Applications to Colloidal and Biological Systems"*, Academic Press; 2nd edition, January 15, (1992).

[25]. G. Binnig, H. Rohrer, *Helv. Phys. Acta* 55, 726 (1982).

[26]. G. Binnig, C.F. Quate, C. Gerber, *Phys. Rev. Lett.* 56, 933 (1986).

[27]. D. Sarid. *"Scanning Force Microscopy with Applications to Electric Magnetic and Atomic Forces"*, Oxford University Press, Oxford (1994).

[28]. G. Meyer and N.M. *Amer. Appl. Phys. Lett.* 53, 1045 (1988).

[29]. G.P. Zhang, Zong, X.F. and George, T.F., *J. Chem. Phys.* 110, 9765 (1999).

[30]. G.P. Zhang, *Phys. Rev. B* 60, 11482 (1999).

[31]. G.P. Zhang and Zong, X.F., *Chem. Phys. Lett.* 308, 289 (1999).

[32]. G.P. Zhang, T.F. George and L.N. Pandey, *J. Chem. Phys.* 109, 2562 (1998).

[33]. Z. Gamba, *J. Chem. Phys*, 97, 1, 553 (1992).

[34]. W.A. Ducker, T.J. Senden, and R.M. Pashley, *Nature*, 353: p. 239-241 (1991).

[35]. G.U. Lee, L.A. Chrisey, and R.J. Colton, Direct measurement of the forces between complementary strands of DNA. Science (USA), 266,(5186): p. 771-3 (1994).

[36]. W.A. Ducker, T.J. Senden, R.M. Pashley, *Langmuir*, 2, 1831 (1992).

[37]. H. Sekiguchi, H. Arakawa, T. Okajima and A. Ikai, *Applied Surface Science*, 188, 3-4, 28, 489, March (2002).

[38]. G.A. Mansoori, "*Advances in atomic & molecular nanotechnology*" in *Nanotechnology, United Nations Tech Monitor*, 53, Sep-Oct (2002).

[39]. A.T. Andrews. "*Electrophoresis*" Oxford University Press; (June 1992); J. Hunter, (editor), "Zeta Potential in Colloid Science". Academic Press, NY (1981).

[40]. G.A. Mansoori, L. Assoufid, T.F. George, G. Zhang, "Measurement, Simulation and Prediction of Intermolecular Interactions and Structural Characterization of Organic Nanostructures" in Proceed. of Conference on Nanodevices and Systems, Nanotech 2003, San Francisco, CA, February 23-27, (2003).

[41]. P. Honenberg and W. Kohn. *Physical Review*, 136:864B, (1964).

[42]. A. Szabo and N.S. Ostlund, "*Modern Quantum Chemistry*", Dover Publications, INC. Mineola, New York (1996).

[43]. J.L. Hutter and J. Bechhoefer, *J. Appl. Phys.* 73, 4123 (1993); *ibid*, J. Vac. Sci. Technol. B, 12, 2251 (1994).

[44]. B.V. Derjaguin, Y.I. Rabinovich and N.V. Churaev, Nature, 272, 313 (1978).

[45]. A. Diehl, M.C. Babosa and Y. Levin, Europhys.Lett.,53 (1), 86 (2001).

[46]. T. Hugel and M. Seitz, Macromol. Rapid Commun. 22, 13, 989 (2001).

[47]. S.I. Zanette, A.O. Caride, V.B. Nunes, G.L. Klimchitskaya, F.L. Freire, Jr. and R. Prioli, Surface Science, 453, 1-3, 10, 75 (2000).

[48]. R. Wiesendanger. Scanning Probe Microscopy and Spectroscopy, Cambridge University Press, Cambridge (1994).

[49]. E.V. Blagov, G.L. Klimchitskaya, A.A. Lobashov and V.M. Mostepanenko. Surf. Sci. 349, 196 (1996).

[50]. E.V. Blagov, G.L. Klimchitskaya and V.M. Mostepanenko. Surf. Sci. 410, 158 (1998).

[51]. Yu.N. Moiseev, V.M. Mostepanenko, V.I. Panov and I.Yu. Sokolov. Phys. Lett. A 132, 354 (1988).

[52]. Yu.N. Moiseev, V.M. Mostepanenko, V.I. Panov and I.Yu. Sokolov. Sov. Phys. Tech. Phys. (USA) 35, 84 (1990).

[53]. M. Bordag, G.L. Klimchitskaya and V.M. Mostepanenko. Surf. Sci. 328, 129 (1995).

[54]. U. Hartman. Phys. Rev. B 43, 2404 (1991).

[55]. C. Argento and R.H. French. J. Appl. Phys. 80, 6081 (1996).

[56]. W. Baumgartner, P. Hinterdorfer and H. Schindler, Ultramicroscopy, 82, 1-4, 85, (2000).

[57]. K. Schilcher, Ph.D. Thesis, University of Linz, (1997).

[58]. P. Hinterdorfer, W. Baumgartner, H.J. Gruber, K. Schilcher and H. Schindler. Proc. Natl. Acad. Sci. USA 93, 3477 (1996).

[59]. P. Hinterdorfer, K. Schilcher, W. Baumgartner, H.J. Gruber and H. Schindler. Nanobiology 4, 177 (1998).

[60]. O.H. Willemsen, M.M.E. Snel, K.O. Werf, B.G. Grooth, J. Greve, P. Hinterdorfer, H.J. Gruber, H. Schindler, Y. Kooyk and C.G. Figdor, Biophys. J. 75, 2220 (1998).

[61]. U. Dammer, M. Hegner, D. Anselmetti, P. Wagner, M. Dreier, W. Huber and H.-J. Güntherodt Biophys. J. 70, 2437 (1996).

[62]. K.A.N.A. Wadu-Mesthrige and G.Y. Liu, Scanning 22, 380 (2000).

[63]. P. Blaha, Schwarz, K., Sorantin, P. and Trickey, S.B., *Comput. Phys. Commun.* 59, 399 (1990).

[64]. G.P. Zhang, Woods, G.T., Shirley, Eric L., Callcott, T.A., Lin, L., Chang, G.S., Sales, B.C., Mandrus, D. and He, J. *Phys. Rev. B* 65, 165107 (2002); Zhang, G.P., Callcott, T.A., Woods, G.T., Lin, L., Sales, Brian, Mandrus, D. and He, J. *Phys. Rev. Lett.* 88, 077401 (2002); 88, 189902 (Erratum) (2002).

[65]. R. Ru, Singh, D. and Krakauer, H., *Phys. Rev. B* 43, 6411 (1991).

[66]. B. Kohler, Wilke, S., Scheffler, M., Kouba, R. and Ambrosch-Draxl, C., *Comp. Phys. Commun.* 94, 31 (1996).

[67]. J.M. Soler and Williams, A.R., *Phys. Rev. B* 40, 1560 (1989).

[68]. H.G. Krimmel, Ehmann, J., Elsässer, C., Fähnle, M. and Soler, J.M., *Phys. Rev. B* 50, 8846 (1994).

[69]. H. Rafii-Tabar, Physics Reports, 325, 239 (2000).

[70]. H. Rafii-Tabar, Physics Reports, 365. 145 (2002).

[71]. S. Erkoc, Physics Reports 278, 79 (1997).

[72]. Frankcombe, T.J., Stranger, R., Schranz, H.W., "The intermolecular potential energy surface of CO_2-Ar and its effect on collisional energy transfer", Internet Journal of Chemistry, 1, 12 (1998).

[73]. D.W. Brenner, Phys. Stat. Sol. (b) 271, 23 (2000).

[74]. F. Ercolessi, M. Parrinello and E. Tosatti, Philos. Mag. A 58, 213 (1988).

[75]. H. Cox, R.L. Johnston and J.N. Murrell, J. Sol. State Chem., 517, 145 (1999).

[76]. M.S. Daw and M.I. Baskes, Phys. Rev. Lett. 50, 1285 (1983).

[77]. M.S. Daw and M.I. Baskes, Phys. Rev. B 29, 6443 (1984).

[78]. M.S. Daw, S.M. Foiles and M.I. Baskes, Mater. Sci. Rep. 9, 251 (1993).

[79]. S.M. Foiles, M.I. Baskes and M.S. Daw, Phys Rev. B 33, 7983 (1986).

[80]. S.M. Foiles and M.S. Daw, J. Mater. Res. 2, 5 (1987).

[81]. R.A. Johnson, Phys. Rev. B 37, 3924 (1988).

[82]. J.B. Adams, S.M. Foiles and W.G. Wolfer, J. Mater. Res. 4, 102 (1989).

[83]. R.A. Johnson, Phys Rev. B 39, 12554 (1989).

[84]. R.A. Johnson, Phys. Rev. B 41, 9717 (1990).

[85]. F. Ercolessi, E. Tosatti and M. Parrinello, Phys. Rev. Lett. 57, 719 (1986).

[86]. M.W. Finnis and J.E. Sinclair, Philos. Mag. A 50, 45 (1984).

[87]. G.J. Ackland and V. Vitek, Phys. Rev. B 41, 19324 (1990).

[88]. A.P. Sutton and J. Chen, Philos. Mag. Lett. Vol. 61, 3, 139 (1990).

[89]. H. Rafii-Tabar and A.P. Sutton, Philos. Mag. Lett. 63, 217 (1991).

[90]. J.E. Black, Phys. Rev. B 46, 4292 (1992).

[91]. Yi-G. Zhang and G.J. Guo, Phys. Earth. Planet. Inte. 122, 289 (2000).

[92]. A. Ghazali and J.-C.S. Levy, Surf. Sci. 486, 33 (2001).

[93]. P. Hohenberg and W. Kohn, Phys. Rev. B136, 864 (1964).

[94]. M.J. Stott and E. Zaremba, Phys. Rev. B22, 1564 (1980).

[95]. J.K. Norskov and N.D. Lang, Phys. Rev. B 21, 2131 (1980).

[96]. A.F. Voter , in: H. Westbrook and R.L. Fleischer(Eds.), Intermetallic Compounds: Vol 1, Principles,John Wiler and Sons Ltd , London, 77 (1994),.

[97]. S.M. Foiles, Phys. Rev. B 32, 7685 (1985).

[98]. M.I. Baskes, Phys. Rev. Lett. 59, 2666 (1987).

[99]. S. Chantasiriwan and F. Milstein, Phys. Rev. B 48, 14080 (1996).

[100]. S. Chantasiriwan and F. Milstein, Phys. Rev. B 58, 5996 (1998).

[101]. G.J. Ackland, G. Tichy, V. Vitek and M.W. Finnis, Philos. Mag. A 56, 735 (1987).

[102]. W.A. Harrison, Electronic Structure and Properties of Solids, Freeman, San Francisco, (1984).

[103]. A.E. Carlsson, Solid State Physics, 43, 1 (1990).

[104]. J. Uppenbrink and D.J. Wales, J.Chem.Phys. 96, 8520 (1992).

[105]. B.R. Eggen, R.L. Johnston, S. Li and J.N. Murrel, Mol. Phys., 76, 619 (1992).

[106]. L.D. Lloyd and R.L. Johnston, Chem. Phy.236, 107 (1998).

[107]. B.S. Wherrett, Group Theory for Atoms, Molecules and Solids, Prentice-Hall International (1986).

[108]. D.G. Pettifor, J. Phys. C 3, 367 (1970).

[109]. W. Harrison, "Elementary Electronic Structure" World Scientific Pub. Co., River Edge, NJ (1999).

[110]. G.A. Mansoori, C. Jedrzejek, N.H. Shah, M. Blander, "Chemical Metallurgy - A Tribute to Carl Wagner" N.A. Gokcen (ed.), The Metallurgical Society of AIME 233-240, (1981).

[111]. J.A. Moriarty, Phys. Rev. B 42, 1609 (1990).

[112]. J.A. Moriarty, Phys. Rev. B 49, 12431 (1994).

[113]. G.C. Abell, Phys. Rev. B 31, 6184 (1985).

[114]. J. Tersoff, Phys. Rev. Lett. 56, 632 (1986).

[115]. J. Tersoff, Phys. Rev. B 37, 6991 (1988).

[116]. J. Tersoff, Phys. Rev. B 38, 9902 (1988).

[117]. J. Tersoff, Phys. Rev. Lett. 61, 2879 (1988).

[118]. J. Tersoff, Phys. Rev. B 39, 5566 (1989).

[119]. D.W. Brenner, Phys. Rev. B 42, 9458. 30 (1990).

[120]. D.W. Brenner, O.A. Shenderova, J.A. Harrison, S.J. Stuart, B. Ni and S.B. Sinnott and J.A. Harrison, J. Phys.: Condens. Matter 14, 783 (2002).

[121]. D. Dharamvir and V.K. Jindal, Int. J. Mod. Phys. B 6, 281 (1992).

[122]. R.S. Ruoff and A.P. Hickman, J. Phys. Chem. 97, 2494 (1993).

[123]. A. Cheng and M.L. Klein, Phys. Rev. B 45, 1889 (1992).

[124]. W.A. Steele, The Interaction of Gases with Solid Surfaces, Pergamon Press, New York (1974).

[125]. L.A. Girifalco, J. Chem. Phys. 96, 858 (1992).

[126]. H. Rafii-Tabar, H. Kamiyama and M. Cross, Surf. Sci. 385, 187 (1997).

[127]. D.J. Oh, R.A. Johnson, Mat. Res. Soc. Symp. Proc. 141, 51 (1989).

[128]. F. Milstein, J. App. Phys. 44, 3825 (1973).

[129]. M. Born and K. Huang, Dynamical Theory of Crystal Lattices, Clarendon Press, Oxford (1954).

[130]. N. Nishioka, T. Taka and K. Hata, Philos. Mag. A 65, 227 (1992).

[131]. D. Srolovitz, K. Maeda, V. Vitek and T. Egami, Philos. Mag. A 44, 847 (1981).

[132]. Axilrod, B.M., J. Chem. Phys. 19, 719, 729 (1951).

[133]. R.M. Lynden-Bell, J. Phys.: Condense Matter 7, 4603 (1995).

[134]. M.P. Allen and D.J. Tildesley, Computer Simulation of Liquids, Clarendon Press, Oxford, UK (1987).

Chapter 3

Thermodynamics and Statistical Mechanics of Small Systems

"*I am of the opinion that the task of theory consists in constructing a picture of the external world that exists purely internally and must be our guiding star in all thought and experiment; that is in completing, as it were, the thinking process and carrying out globally what on a small scale occurs within us whenever we form an idea.*"

The immediate elaboration and constant perfection of this picture is then the chief task of theory. Imagination is always its cradle, and observant understanding its tutor. How childlike were the first theories of the universe, from Pythagoras and Plato until Hegel and Schelling. The imagination at that time was over-productive, the text by experiment was lacking. No wonder that these theories became the laughing stock of empiricists and practical men, and yet they already contained the seeds of all the great theories of later times: those of Copernicus, atomism, the mechanical theory of weightless media, Darwinism and so on." **Ludwig Boltzmann** (in *"On the Significance of Theories"* 1890).

Introduction

This chapter is concerned with one of the important subjects of computational nanotechnology, namely thermodynamics and statistical mechanics and their applications in molecular systems to predict the properties and performances involving nanoscale structures. A scientific and technological revolution has begun in our ability to systematically organize and manipulate matter on a bottom-up fashion starting from atomic level as well as design tools, machinery and energy conversion devices in nanoscale towards the development of nanotechnology. There is also a parallel miniaturization activity to scale down large tools, machinery and energy conversion systems to micro and nanoscales towards the same goals [1-4]. The science of miniaturization is being developed and the limitations in scaling down large systems to nanoscale are under investigation. Advancement and knowledge about the thermodynamics of small systems is essential to help achieve these goals.

Principles of thermodynamics and statistical mechanics for macroscopic systems are well defined and mathematical relations between thermodynamic properties and molecular characteristics are derived. The objective here is to introduce the basics of the thermodynamics of small systems and introduce statistical mechanical techniques, which are applicable to small systems. This will help to link the foundation of molecular based study of matter and the basis for nano science and technology. Development of predictive computational and modeling options for the behavior of nano systems will depend on advancements in their thermodynamics.

The subject of thermodynamics of small systems was first introduced by T.L. Hill in two volumes in 1963 an 1964 [5] to deal with chemical thermodynamics of mixtures, colloidal particles, polymers and macromolecules. Nanothermodynamics, a term which is recently introduced in the literature by T.L. Hill [6-8], is a revisitation of the original work of T.L. Hill mentioned above on thermodynamics of small systems.

Significant accomplishments in performance and changes of manufacturing paradigms are only possible with the advancement of principles of science of small systems including thermodynamics. The answer to the question of how soon will this next industrial revolution arrive depends, a great deal, on the intensity of such scientific activities.

Thermodynamics, in general, is concerned with those physical and chemical phenomena, which involve heat and temperature. From the practical definition, thermodynamics is the phenomenological theory of converting heat to work and understanding the role of energy and other properties of matter in this conversion process. Equilibrium thermodynamics is confined to the equilibrium states of matter. Change of one equilibrium state to another is infinitely slow and as a result it is independent of time. Concepts of heat, work, energy, and properties of matter exist in public vocabulary of every language. Thermodynamic laws, which govern the relations between these concepts, originate from ordinary experiences in our daily lives. Statistical thermodynamics, in general, is concerned with the atomic or molecular structure of matter and the relationship between microscopic structure and the macroscopic thermodynamic behavior of substances.

Historically, thermodynamics took its name from the study of efficiency of heat engines by Carnot. In his 1824 thesis, Carnot stated that the efficiency of a heat engine depended only on the temperature difference between its heat source and heat sink and not on the working substance. A decade was passed when Clapeyron developed the relationship between vapor pressure and an unknown function of empirical temperature scale. Clausius later identified this unknown function as the absolute temperature scale. The equation relating the vapor pressure to the absolute temperature is known as the Clausius-Calpeyron equation [9]. It was in 1850 when Clausius published his thesis on the Second Law of thermodynamics, which is known as "the Clausius statement." It was this statement that marked the beginning of thermodynamics as a science as it was stated by Gibbs in the late nineteenth century. A year after publication of this thesis, Thomson formulated explicitly the First and Second Laws of thermodynamics. Thomson had already defined an absolute temperature scale in 1848 and was aware of the 1845 publication by Joule in which Joule had demonstrated the equivalence of heat and work.

It was 1865 when Clausius introduced the term "entropy" and stated "The energy of universe is constant. The entropy of universe tends towards a maximum". Introduction of the term entropy resulted in a new formulation of the Second Law by Clausius.

Since the findings of Clausius, many other investigators have contributed to the science of thermodynamics. The thesis by Gibbs published in 1875 entitled "On the equilibrium of heterogeneous substances" and his other publications have a special place in thermodynamics of mixtures and phase equilibrium. Gibbs extended the science of thermodynamics in a general form to heterogeneous systems with and without chemical reactions. He is also credited for derivation and formulation of completely general equilibrium conditions for various cases. Other early contributors to this branch of thermodynamics are the following: (i) Helmholtz, who in 1882 independent of Gibbs, introduced the concept of free energy and derived the relationship now known as the Gibbs-Helmholtz equation, (ii) Duhem, who in 1886, derived the Gibbs-Duhem equation, Planck, who in 1887, divided the changes of state into two classes of thermodynamic processes, namely reversible and irreversible processes, (iii) Nerst, who in 1906, published his heat theorem, and Carathe'odory, who in 1909, developed a new axiomatic basis of thermodynamics.

In the present chapter, we also introduce the basics of thermodynamics for small systems, which is of interest in the field of nanotechnology, miniaturization and the science of conversion of thermal to mechanical energy in small systems including the laws of thermodynamics of nano systems. It must be emphasized that what is generally known as thermodynamics in the literature deals with systems in equilibrium. However, the materials presented in this chapter are general and are not limited to systems in equilibrium.

Thermodynamic Systems in Nanoscale

The definition of a thermodynamic system in nanoscale is quite similar to that of macroscopic systems. In thermodynamics, a system is any region completely enclosed within a well-defined boundary. Everything outside the system is then defined as the surroundings. The boundary may be either rigid or movable. It can be impermeable or it can allow heat, work or mass to be transported through it. In any given situation a system may be defined in several ways.

Although it is possible to speak of the subject matter of thermodynamics in a general sense, the establishment of any relationships among thermodynamic properties of a system requires that they be related to a particular system. The known macroscopic thermodynamic property relations apply to systems, which contain a very large number of atoms or molecules. The validity of such relations to nano systems is open to question.

The state of a system is an important concept in thermodynamics and is defined as the complete set of all its properties, which can change during various specified processes. The properties, which comprise this set, depend on the kinds of interactions, which can take place both within the system and between the system and its surroundings. Any two systems, subject to the same group of processes, which have the same values of all properties in this set, are then indistinguishable and we describe them as being in identical states.

A process in thermodynamics is defined as a method of operation in which specific quantities of heat and various types of mass and work are transferred to or from the system to alter its state. One of the objectives of thermodynamics is to relate these state changes in a system to the quantity of energy in the form of heat and work transferred across its boundaries.

The simplest system in nanoscale may be chosen as a single particle, like an atom or molecule, in a closed space with rigid boundaries. In the absence of chemical reactions, the only processes in which it can participate are transfers of kinetic or potential energy to or from the particle, from or to the walls. The state for this one-particle system is a set of coordinates in a multi-dimensional space indicating its position and its momentum in various vector directions. For example, a simple rigid spherical monatomic molecule would require six such coordinates, three for its position and three for its momentum in order to completely define its state. For a system containing N rigid spherical monatomic molecules the state of the system is defined by the 6N set of all position and momentum values for all the particles. They can thus determine all the thermodynamic properties of the group. For systems consisting of very large number of such particles in equilibrium, the thermodynamic property relations are already established (see for example MCSL theory [10]). For systems containing limited number of rigid sphere particles, the thermodynamic property relationships need to be established.

The set of all the thermodynamic properties of a multi-particle system including its temperature, pressure, volume, internal energy, etc is defined as the thermodynamic state of this system. An important aspect of the relationships between thermodynamic properties in a large, macroscopic and also known as extensive system, is the question of how many different thermodynamic properties of a given system are independently variable. The number of these represents the smallest number of properties, which must be specified in order to completely determine the entire thermodynamic state of the system. All other thermodynamic properties of this system are then fixed and can be calculated from these specified values. The number of these values, which must be specified, is called the variance or the degrees of freedom of the system. Small or nano systems, on the other hand, could be looked at from the point of view of nonextensivitiy or thermodynamics of nonextensive systems. This subject will be discussed in this chapter as well.

The systems of interest to nanoscience and nanotechnology are the isolated individual nanostructures and their assemblies, small droplets and bubbles, clusters, and fluids and solids confined in small spaces, like inside nanotubes and fullerene, and going through phase changes. Nano systems are made of countable (limited) number of atoms or molecules. Their sizes are larger than individual molecules and smaller than micro systems. One of the characteristic features of nano systems is their high surface-to-volume ratio. Their electronic and magnetic properties are often distinguished by quantum mechanical behavior. Their mechanical and thermal properties can be formulated within the framework of classical statistical mechanics of small systems as presented by Hill [5-8], and through the newly developed nonextensive statistical mechanics as presented in this Chapter.

Small systems of interest in nanotechnology are generally made up of condensed (liquid or solid) matter. Nano systems can contain all forms of condensed matter, be it soft or hard, organic or inorganic and/or biological components. A deep understanding of the interactions between individual atoms and molecules composing the nano systems and their statistical mechanical modeling is essential for nanotechnology.

Energy, Heat and Work in Nanosystems

Energy is a quantity which measures a combination of effort expended and results produced. For macroscopic systems energy is often defined as "the ability to perform work" by a system under consideration. At the macroscopic level, the idea expressed is probably sufficient in most cases, but for nano systems this definition is not sufficient enough because the term "work" requires a more precise definition than it's generally well understood macroscopic definition. A useful definition of energy from the viewpoint of thermodynamics of small systems could be "the capacity to induce a change which would be visible in nanoscale". This is to distinguish work from heat, which is a change in the energy of the system with no visible change in nanoscale. It must be pointed out that this definition is also subject to error as the size of a nano system decreases to a level that its fluctuations are of larger scale and closer to molecular motions.

Work is generally defined as the product of a driving force and its associated displacement, which represents a means of transfer of energy between a system and its surroundings. A driving force causes and controls the direction of change in energy, which is a property of the system. The quantitative value of this change is called a "displacement". In thermodynamics, this quantity has meaning only in relation to a specifically defined system. Relative to a particular system there are generally two ways of locating a driving force and the displacement it produces.

In one way both the driving force and the displacement are properties of the system and are located entirely within it, so that the energy calculated from their product represents a change in the internal energy of the system.

In another way, however, the displacement could occur within the system but the driving force producing it is a property of the surroundings and is applied externally at the system. Similarly, both the driving force and its displacement could be located entirely within the surroundings. In both of these cases the calculated energy is then a change in the total energy of the system. The total energy of a system is the sum of its internal, kinetic and potential energies.

Heat is another means of exchange of energy between the system and its surroundings through the system boundary. The distinction between heat and work is the fact that heat is a means of energy with no visible change in nanoscale but in random atomic and molecular translation, rotation and vibrations on the boundary of the system.

By definition, the boundary of a system is a region of zero thickness containing no matter at all. In any quantitative application of thermodynamics, it is always important to make a careful distinction between energy changes within a system and the concepts of work and heat, which are means of exchange of energy between the system and its surroundings.

In what follows we introduce the specificities of the thermodynamics of nano systems, which includes an examination of laws of thermodynamics and the concept of nonextensivity of nano systems.

Laws of Thermodynamics

Classical thermodynamics is a typical example of an axiomatic science. The rigorous deduction of thermodynamic concepts from its basic axioms is an intricate logical problem [11-13]. In an axiomatic science certain aspects of nature are explained and predicted by deduction from a few basic axioms which are assumed to be always true. The axioms themselves need not be proved but they should be sufficiently self-evident to be readily acceptable and certainly without known contradictions. The application of thermodynamics of large systems to the prediction of changes in given properties of matter in relation to energy transfers across its boundaries is based on four fundamental axioms, the Zero[th], First, Second, and Third Laws of thermodynamics. The question whether these four axioms are necessary and sufficient for all systems whether small or large, including nano systems, is something to be investigated in the future. In the present discussion, we examine the format of these four axioms for nano systems.

Thermodynamics and statistical mechanics of large systems consisting of many particles beyond the thermodynamic limit (V and N go to infinity, $N/V = \rho_N$ is finite) is well-developed [14-16]. Normally the science of thermodynamics describes the *most likely macroscopic behavior* of large systems (consisting of 10^{23} particles ~ *1 ccm* in volume or more) under slow changes of a few macroscopic properties. The really large systems (like astrophysical objects) as well as small systems (like those of interest in nanotechnology) are excluded.

Recent developments in nanoscience and nanotechnology have brought about a great deal of interest into the extension of thermodynamics and statistical mechanics to small systems consisting of countable particles below the thermodynamic limit. Hence, if we like to extend thermodynamics and statistical mechanics to small systems in order to remain on a firm basis we must go back to its founders and, like them establish new formalism of thermodynamics and statistical mechanics of small systems starting from the safe grounds of mechanics. One should point out difficulties with and possibly the changes that we have to introduce in thermodynamics and statistical mechanics to make them applicable to small systems.

Structural characteristics in nanoscale systems are dynamic, not the static equilibrium of macroscopic phases. Coexistence of phases is expected to occur over bands of temperature and pressure, rather than along just sharp points. The pressure in a nano system cannot be considered isotropic any more and must be generally treated as a tensor. The Gibbs phase rule looses its meaning, and many phase-like forms may occur for nanoscale systems that are unobservable in the macroscopic counterparts of those systems [5,17].

Such questions as property relations and phase transitions in small (nano) systems are subjects to be investigated and formulated provided the formulation of working equations of thermodynamics and statistical mechanics of small systems are developed. The principles of phase transitions in small systems are presented in Chapter 7 of this book. One distinction between macroscopic and small systems is the fact that while the thermodynamic property relations (equations of state) of macroscopic systems are independent of the environment, they are environment-dependent in small systems [5].

It is worth mentioning that the molecular self-assembly (bottom-up technology) which is originally proposed by Feynman [18] has its roots in phase transitions. As Hill has pointed out [5] small system effects will be particularly noticeable at phase transitions and in the critical region. A comprehensive understanding of thermodynamics and statistical mechanics of small systems will help bring about the possibility of formulation of alternative self-assemblies of various atomic and molecular assemblies. The subjects of phase transitions and self-assemblies in small systems will be discussed in Chapters 7 and 9, respectively.

To formulate the thermodynamics of small systems, one has to start evaluating thermodynamics from the first principles reviewing the concepts, laws, definitions and formulations and draw a set of guidelines for their applications to small systems. Historically thermodynamics was developed for understanding the phenomena of converting thermal energy to mechanical energy. With the advent of technology thermodynamics assumed a universal aspect applicable for all kinds of systems, whether open or closed, and under variations of temperature, pressure, volume and composition. The concept of irreversible thermodynamics [19,20] is an extension of thermodynamics to dynamic systems under various external forces generating varieties of fluxes in the system. It is quite appropriate to examine the laws of thermodynamics for a small (nano) system, which does not satisfy the "thermodynamic limit".

The Zero[th] Law

The Zero[th] Law of thermodynamics consists of the establishment of an absolute temperature scale. Temperature of a macroscopic system is a well-defined property, its fluctuations are quite negligible, and it is a measure of its thermal equilibrium (or lack of it). While the temperature of a small system is also a well-defined thermodynamic property as is in macroscopic systems, it generally has larger fluctuations with respect to time and space which would magnify as the size of the system reduces.

In nano systems, as it is well explained in the following statement by the US National Initiative on Nanotechnology [2] fluctuations play an important role: *"There are also many different types of time scales, ranging from 10^{-15} s to several seconds, so consideration must be given to the fact that the particles are actually undergoing fluctuations in time and to the fact that there are uneven size distributions. To provide reliable results, researchers must also consider the relative accuracy appropriate for the space and time scales that are required; however, the cost of accuracy can be high. The temporal scale goes linearly in the number of particles N, the spatial scale goes as O(N.logN), yet the accuracy scale can go as high as N^7 to N! with a significant prefactor."* While fluctuations in small systems are recognized to be generally appreciable in magnitude, they are not yet quantitatively correlated to their properties. If we are going to extend applications of thermodynamics and statistical mechanics to small systems, accurate calculation of fluctuations are quite necessary.

It must be pointed out that, in addition to temperature fluctuations, appreciable fluctuations may also exist in pressure and energies (internal energy, Gibbs free energy and Helmholtz free energy) of small systems. Equilibrium is defined as the condition when there is no visible change in the temperature, pressure and energies of a system with respect to time. Due to the appreciable fluctuations in small systems, there will be difficulties in defining the concepts of thermal, mechanical and chemical equilibrium.

Another important concept in thermodynamics is reversibility and irreversibility. A reversible system is one that when the direction of a driving force (like heat or work) is changed, the system goes back to its original condition. Thermodynamic reversibility in macroscopic systems has become synonymous with thermodynamic equilibrium. A nano system, on the other hand, may be reversible, but it may not be at equilibrium. Similar to macro systems a small system can be defined as isolated, closed (also known as control mass), open (also known as control volume), adiabatic, isothermal, isometric (or isochoric, also known as constant volume) and isobaric (constant pressure).

It is customary in statistical mechanics to express fluctuations in thermodynamic properties in terms of distribution functions and, thus, as derivatives of other properties [5,16]. Recent advances in the fluctuation theory of statistical mechanics of macroscopic systems has been quite successful to quantify and predict the behavior of complex mixtures [16] for which intermolecular potential energy functions are not accurately available.

The First Law

The First Law of thermodynamics as defined for macroscopic systems in which no nuclear reactions is taking place is simply the law of conservation of energy and conservation of mass. When, due to nuclear reactions, mass and energy are mutually interchangeable, conservation of mass and conservation of energy should be combined into a single conservation law which, as far as we know, is universal.

The intuitive acceptability of this law for all systems, whether in nano or macroscopic scales, is apparent. The concepts of conservation of mass and energy in various forms have existed from antiquity, long before any precise demonstration of them could be made. The verse in Galatians 7, of the Book of Judges "whatever a man sows, this he will also reap." is, in a sense, the biblical affirmation of the law of conservation of mass. The Greek philosopher Empedocles developed a theory of physical elements. He believed that everything was made up of four elements: earth, fire, air, water. He considered matter indestructible, although its forms--earth, fire, air, or water-- could be interchanged. The situation was confused in the middle Ages by a feeling that fire actually "destroyed" the matter, which burned. This was set right in 1774 when the French chemist, Antoine Laurent Lavoisier, conclusively demonstrated the conservation of mass in chemical reactions.

An intuitive feeling for energy conservation is also deep-rooted even though its demonstration is experimentally more difficult than that for mass conservation. The earliest demonstration of energy conservation made in 1797 is credited to Count Rumford, known primarily for the work he did on the nature of heat and for whom the Rumford fireplace is named. Fortunately, he made his observations in a cannon factory where the mechanical work of boring cannon was converted into heat and transferred to a large amount of cooling water. If this heat transfer had not been very large relative to the heat losses, the correct conclusion could easily have escaped him. The first accurate measurement and indisputable demonstration of the precise equivalence of mechanical work and the total thermal energy obtainable from it did not occur until the experiment of Joule in 1843.

In 1843 Joule developed an experiment to convert the mechanical energy to thermal energy through fluid motion and mixing of water, known as the Joule's water friction experiment. A well-insulated calorimeter was equipped with baffles to increase the frictional drag on the paddles. A weight falling a measured distance performs a known amount of work. By measuring the temperature increase in the water in the calorimeter as the weight falls, the amount of heat generated by the work can be calculated. Through the Joule's experiment, we learned about the equivalence and relationship of thermal energy and the mechanical energy spent to generate it. Later, the metric unit of energy was named in honor of Joule (1 Calorie = 4.186 Joule).

In another experiment, Joule measured the heat produced by compression of air and by electrical currents. In 1840, he stated a law, now called Joule's Law, that heat is produced in an electrical conductor. Together with W. Thomson (Lord Kelvin) he discovered in 1853 the effect known as Joule-Thomson effect. Discovery of the principle of magnetostriction is also credited to Joule. He did not claim, however, to have formulated a general Law of Conservation of Energy. Nevertheless, his experiments were certainly fundamental in bringing that formulation about [21].

The First Law of thermodynamics, which is the same as the principle of conservation of mass and energy, provides us with the relation between heat, work and energy content of a system. The equation for the First Law of thermodynamics for a closed (controlled mass) macro system takes the following form,

$$dE = \delta Q_{in} + \delta W_{in} \qquad (1)$$

where dE stands for the exact differential energy increase in the system due to addition of inexact differential amounts of heat and work, δQ_{in} and

$$\delta W_{in} = -\sum_i P_{ii}^{ext} d\varepsilon_{ii} V \,,$$

respectively, to the system, P_{ii}^{ext} (i=x,y,z) are the components of the external diagonal pressure tensor $\vec{P}^{ext}$ and $d\varepsilon_{ii}$ is the specific deformation of volume V due to the external pressure tensor [22]. This equation is valid for small systems as well as large systems. However, clear distinction / separation between the terms heat and work may be a bit foggy for certain small systems and especially when the size of the nano system decreases.

As we mentioned before a useful definition of energy from the viewpoint of thermodynamics of small systems could be "the capacity to induce a change, which would be visible in nanoscale". There are three kinds of energy important in thermodynamics: Kinetic energy, potential energy, and internal energy due to intermolecular interactions in a system. Work and heat are means of energy exchanges between a system and its surroundings· or another system. Transfer of energy through work mode is a visible phenomenon in macroscopic systems. However, it is invisible in a nano system, but it occurs as a result of the collective motion of an assembly of particles of the nano system resulting in changes in energy levels of its constituting particles. Transfer of energy through heat mode is also an invisible phenomenon, which occurs in atomic and molecular level. It is caused by a change not of the energy levels but of the population of these levels.

While the First Law, which is actually the conservation of energy, is one of the fundamental laws of physics, it alone cannot explain everything related to conversion of thermal to mechanical energy. Another law is required to understand that the thermal energy can change form to mechanical energy. For instance, how the energy in a barrel of oil can move an automobile and for what maximum distance. This is the realm of the Second Law providing the inequality expression for the entropy production as is discussed below.

The Second Law

Lord Kelvin originally proposed the Second Law of thermodynamics in the nineteenth century. He stated that heat always flows from hot to cold. Rudolph Claussius later stated that it was impossible to convert all the energy content of a system completely to work since some heat is always released to the surroundings. Kelvin and Claussius had macro systems in mind where fluctuations from average values are insignificant in large time scales. According to the Second Law of thermodynamics for a closed (controlled mass) system we have [20],

$$dP_S = dS - \delta Q_{in}/T_{ext} \geq 0. \tag{2}$$

This definition of the differential entropy production in a closed (controlled mass) system, dP_S, which is originally developed for macro systems is still valid for nano / small systems. As it will be demonstrated later, for small systems the term entropy, S, may assume a new statistical form due to nonextensivity of nano systems.

Now, by joining the First and the second Law equations for a closed system and with the consideration of the definition of work,

$$\delta W_{in} = -\sum_i P_{ii}^{ext} d\varepsilon_{ii} V \,,$$

the following inequality can be derived for the differential total entropy of a system:

$$dS \geq \frac{dE}{T_{ext}} + \frac{1}{T_{ext}} \sum_i P_{ii}^{ext} d\varepsilon_{ii} V \tag{3}$$

This is a general inequality, which seems to be valid for large as well as small systems. Of course, for a system in equilibrium the inequality sign will be removed and the resulting equation is the well-known Gibbs equation. However, due to appreciable fluctuations in properties of small systems, definition of the static equilibrium, as we know it for large systems, will be difficult, if not impossible to make. In nano systems and in very short periods of time such fluctuations are observed to be significant, violating the Second Law [23]. However, for longer times nano systems are expected to be closer to reversibility than macroscopic systems.

The Third Law
The Third Law of thermodynamics for large systems, also known as "the Nernst heat theorem", states that the absolute zero temperature is unattainable. Although one can approach absolute zero, one cannot actually reach this limit. The third law of thermodynamics states that if one could reach absolute zero, all bodies would have the same entropy. In another words, a body at absolute zero could exist in only one possible state, which would possess a definite energy, called the zero-point energy. This state is defined as having zero entropy.

It is impossible to reach absolute zero because properties of all systems are in dynamic equilibrium and their atomic and molecular properties fluctuate about their average values. Likewise, the energy per particle of a macroscopic system is an average value, and individual particle energies of the system will fluctuate around this value. As a result, temperature cannot reach absolute zero due to energy fluctuation.

The Third Law for a macroscopic system also states that the limit of the entropy of a substance is zero as its temperature approaches zero, a concept necessary in making absolute entropy calculations and in establishing the relationship between entropy as obtained from the statistical behavior of a multi-particle system.

The non-attainability of absolute zero temperature seems also to be valid for nano systems due also to fluctuations. However, it may be possible to devise a confined nano system whose fluctuations are more damped. As a result, such a system is more likely to approach closer to absolute zero temperature than macro systems. One could make it to have even negative temperature such as for paramagnetic spins, which are cooled off by applying a magnetic field so that entropy decreases with energy and leads to negative temperatures [24].

Currently, the third law of thermodynamics is stated as a definition: the entropy of a perfect crystal of an element at the absolute zero of temperature is zero. This definition seems to be valid for small systems as well as large systems.

Statistical Mechanics of Small Systems

The objective of statistical mechanics is generally to develop predictive tools for computation of properties and local structure of fluids, solids and phase transitions from the knowledge of the nature of molecules comprising the systems as well as intra- and intermolecular interactions. The accuracy of the predictive tools developed through statistical mechanics will depend on two factors. The accuracy of molecular and intermolecular properties and parameters available for the material in mind and the accuracy of the statistical mechanical theory used for such calculations.

Statistical mechanical prediction of the behavior of matter in macroscopic scale, in the thermodynamic limit (V & $N \rightarrow \infty$ and $N/V = \rho_N$ finite), is well developed and a variety of molecular-based theories and models are available for prediction of the behavior of macroscopic systems. There is also a wealth of data available for thermodynamic and transport properties of matter in macroscopic scale, which can be used for testing and comparison of molecular based theories of matter in macroscopic scale. In the case of nano (small) scale there is little or no such data available and the molecular theories of matter in nanoscale are in their infancy. With the recent advent of tools to observe, study and measure the behavior of matter in nanoscale it is expected that in a near future experimental nanoscale data will become available. In the present section we introduce an analytic statistical mechanical technique which has potential for application in nano systems. In the next chapter, we will introduce the computer simulation techniques, which have been proposed and used for molecular based study of matter in nanoscale.

Statistical mechanics of small systems is not a new subject. Many investigators have studied small systems consisting of one or more macromolecule, droplets, bubbles, clusters, etc. utilizing the techniques of statistical mechanics [4-8,25,26]. Computer simulation approaches, like Monte Carlo and molecular dynamics techniques, have been used extensively for such studies and they will be presented and reviewed in the next few chapters. However, a general analytic formalism for dealing with small systems and developing working equations for thermodynamic properties, without regard to their nature, through statistical mechanics is still lacking.

Recent nanotechnology advances, both bottom-up and top-down approaches, have made it possible to envision complex and advanced systems, processes, reactors, storage tanks, machines and other moving systems which include matter in all possible phases and phase transitions. There is a need to understand and develop analytic predictive models, for example, for the behavior of a matter confined in a fullerene, or flowing in a nanotube at various state conditions of pressures, temperatures and compositions. For such diverse circumstances and for application of such components in the design of nanomachinery, the development of analytic predictive approaches of properties of matter in nanoscale is necessary to build accurate computational techniques to model such nanomachinery.

It is a well-known fact that the behavior of matter confined in nano systems (inside a fullerene, in a nanotube, etc.) is a function of the environmental geometry, size and wall effects that are surrounding it. This has not been the case with macroscopic scale of matter where we have been able to develop universal correlations and equations of state to be applied in all possible applications regardless of the geometry of the confinement systems. For example the property-relations of water (its equation of state), is universally needed for many applications. Such applications include in the use of water as working fluid in thermal to mechanical energy conversion devices, in the use of water as the reagent or solvent in many chemical processes and in oceanographic, meteorological and geothermal studies. However, the same equations of state of water may not be applicable in small systems. Even if one develops such database for a particular nanosystem, the results may not be applicable in other nano system. This demonstrates the need for a methodology for the development of universal analytic techniques of thermodynamic property relations in nanoscale. One of the prime candidates for this endeavor seems to be the newly developed thermodynamics and statistical mechanics of nonextensive systems. In what follows we present the principle of thermodynamics and statistical mechanics of nonextensive systems and discuss its validity for small systems due to their nonextensive nature.

Thermodynamics and Statistical Mechanics of Nonextensive Systems

Statistical mechanical treatment of macroscopic systems consisting of an assembly of molecules starts with the Boltzmann formula along with the use of a statistical ensemble averaging techniques for the probability distribution of molecules / particles in the system. There exist various ensemble-averaging techniques in statistical mechanics including microcanonical, canonical, grand canonical and Gibbs ensembles.

It should be pointed out that the use of thermodynamic limit and of extensivity are closely interwoven with the development of classical statistical mechanics as are reported in many books published on this topic [14,15, 25-30]. Hence, to extend the use of statistical mechanics to small systems we must go back to its foundations and establish the new formalism starting from the appropriate grounds of *mechanics* and *statistics*. One such appropriate approach is the thermodynamics and statistical mechanics of nonextensive systems originally developed in 1988 by Tsallis [31].

In order to explain the nature of nonextensivity of nanoscale systems the following discussion is presented.

In thermodynamics, properties (variables) are classified as being either extensive or intensive. When properties of a system are independent of the number of particles present in the system, they are called "intensive properties (variables)". Otherwise, those properties (variables) are called extensive properties (variables). The test for an intensive property is to observe how it is affected when a given system is combined with some fraction of an exact replica of itself to create a new system differing only in size. Intensive properties are those, which are unchanged by this process. Those properties whose values are increased or decreased in direct proportion to the enlargement or reduction of the system are called "extensive properties". For example, if we exactly double the size of a large or macroscopic system in thermodynamic limit by combining it with an exact replica of itself, all the extensive properties are then exactly double and all intensive properties are unchanged. For nano systems, these rules in defining intensive and extensive thermodynamic properties don't seem, in general, to be satisfied. Accordingly, nano systems are in the category of nonextensive systems.

Euler's Theorem of Homogenous Functions

Euler's theorem of homogeneous functions is used to distinguish the extensive and intensive thermodynamic properties. According to the Euler's theorem of homogeneous functions, a function $f(x_1, x_2, ..., x_r)$ that satisfies the following condition is called a homogeneous function of degree λ:

$$f(tx_1, tx_2, ..., tx_r) = t^\lambda f(x_1, x_2, ..., x_r) \tag{4}$$

For a homogeneous function of degree λ it can be readily shown that the following condition also holds

$$x_1 \frac{\partial f}{\partial x_1} + x_2 \frac{\partial f}{\partial x_2} + \ldots + x_r \frac{\partial f}{\partial x_r} = \lambda f \, . \tag{5}$$

When the Euler's theorem is used for thermodynamic property relations of macroscopic systems, the value of exponent λ defines the nature of thermodynamic variable. When exponent $\lambda = 0$ the thermodynamic variable is intensive and when $\lambda = 1$ the variable will be extensive.

In nano systems distinction of intensive and extensive thermodynamic variables (properties) from one another loses its significance. The definitions of such inherent intensive properties as temperature and pressure also lose their firmness due to fluctuations. Accordingly in small systems we may need to propose a similar relation as the above equation for intensive and extensive properties but with $\lambda \neq 0$ *or* 1. Actually the numerical value of λ may be non-integer and it may vary for systems of varying sizes and natures. This is because; in small systems, we are faced with a new class of thermodynamic properties which do not possess the usual mathematical and physical interpretations of the extensive properties of the macroscopic systems. For determining λ some experimental data or reliable calculations on the relevant properties of the small system under consideration need to be carried out.

Any extensive thermodynamic property is a summation of the same property of the particles of the system, which are related to the energetics and space occupied by the particles. Any intensive thermodynamic property is, in the other hand, the result of an average of the particles' position and momentum coordinates with respect to an external frame of reference.

Consequently, if we alter the number of particles by changing only the size of the system, we should then alter the extensive properties proportionately and retain the intensive properties for large systems in thermodynamic limit. For small / nano systems, this may not be the case, i.e. extensive thermodynamic properties would alter inproportionately and intensive properties may not be retained.

Boltzmann and Boltzmann–Gibbs Formulae of Entropy

In the nineteenth century, Ludwig Boltzmann derived the Second Law by assuming that matter was composed of particulate bodies (atoms, molecules, etc.) by applying Newtonian mechanics along with principles of statistics. According to Boltzmann, the Second Law of thermodynamics is probabilistic in nature. He worked on statistical mechanics using probability to describe how the properties of atoms determine the properties of matter. In particular, he demonstrated the Second Law of thermodynamics in a statistical statement form. According to Boltzmann:

$$S = k_B \ln(W),$$ (6)

where S is the entropy of a system, $k_B = 1.38054 \times 10^{-23}$ *[Joule/Kelvin]* is the thermodynamic unit of measurement of entropy, now known as the Boltzmann constant, W is the "probability" of the system in its mathematical sense, that is the number of distinct ways of arranging the particles consistent with the overall properties of the system.

Two important characteristics of Boltzmann entropy are [30]:

(i) its nondecrease: if no heat enters or leaves a system, its entropy cannot decrease;

(ii) its additivity: the entropy of two systems, taken together, is the sum of their separate entropies.

However, in statistical mechanics of finite (nano) systems, it is impossible to completely satisfy both of the above-mentioned characteristics [32].

Boltzmann entropy, S, as defined by Eq.(6), is for a macroscopic (large) state over a statistical ensemble in equiprobability. It is considered the entropy of a coarse-grained distribution and Gibbs later expressed it when the probabilities are not all the same. In terms of the probability distribution of the observational states of the system resulting in the well-known Boltzmann-Gibbs formula:

$$S = -k_B \sum_{i=1}^{W} p_i \ln p_i.$$ (7)

Gibbs from Eq. (6) derived the Boltzmann-Gibbs expression for entropy in 1870s by considering a system consisting of a large number, N, elements (molecules, organisms, etc.) classified into W classes (energy-states, species, etc.). W is the total number of such microscopic possibilities. In this equation p_i is probability of distribution of a set of particles i in the system. In the case of equiprobability (i.e., $p_i=1/W$) Eq. (7) reduces to Eq. (6), the original Boltzmann formula.

For over a century, engineers, physicists and chemists have used this formula for entropy to describe various macroscopic physical systems. It is the starting point of the science of statistical mechanics and thermodynamics through the formulation of ensemble theories [27-30].

Tsallis Formula of Entropy

Boltzmann and Boltzmann-Gibbs entropy formulas have limitations since they are based on the assumption of additivity of entropy [31]. The very notions of extensivity (additivity) and intensivity in thermodynamics are essentially based on the requirement that the system is homogeneous, which is provided for big systems with weak interactions or, more precisely, in the thermodynamic limit, $(N,V) \rightarrow \infty$, N/V=finite. These notions make no strict sense for inhomogeneous systems such as the small (nano) systems or systems characterized by the size of the order of correlation length [33]. This is indicative of the fact that small / nano systems are truly nonextensive and statistical mechanics of nonextensive systems may be applied for these cases.

It has been demonstrated that the domain of validity of classical thermodynamics and Boltzmann-Gibbs statistics, Eq. (7) is restricted, and as a result, a good deal of attention have been put to discuss such restrictions [31, 33-35]. This branch of science is categorized in a special part of thermodynamics, which is named "nonextensive thermodynamics". Nonextensive thermodynamics or thermodynamics of nonextensive systems has a formalism, which is proper for the study of small as well as other systems that do not exhibit extensivity.

To overcome difficulties of nonextensive systems a new statistics was proposed by Tsallis [31], which has recently been modified [34]. According to Tsallis the entropy can be expressed by the following formula:

$$S_q = k \frac{1 - \sum_{i=1}^{W} p_i^q}{q-1} \quad (\sum_{i=1}^{W} p_i = 1; q \in \Re), \tag{8}$$

where k is a positive constant and W is the total number of microscopic possibilities of the system. The parameter q appearing in this equation is known as the "entropic index".

Equation 6 can be analyzed further depending on the value and sign of entropic index q:

(1) For all positive values of q *(q>0)* since

$$\sum_{i=1}^{W} p_i = 1 \quad \text{and} \quad 0 \leq p_i < 1, \quad \text{then} \quad \sum_{i=1}^{W} p_i^q \leq 1.$$

As a result, it is obvious that S_q as given by Eq. (8) is always positive.

(2) In the case of equiprobability (i.e., $p_i = 1/W$) Eq. 6 reduces to

$$S_q = k \frac{1 - W^{1-q}}{q-1}.$$

Then for $q=1$ this equation reduces to $S_1 = k.\ln W$, which is the same as Eq. (6), the original Boltzmann formula.

(3) In the limit when $q \to 1$ we can write $p_i^{q-1} = e^{(q-1)lnpi} \sim 1 + (q-1) \ln pi$. By replacing this expression in Eq. (8) and in the limit $q \to 1$ we obtain

$$S_1 = -k \sum_{i=1}^{W} p_i \ln p_i,$$

which is the usual Boltzmann-Gibbs formula as given by Eq. (7).

(4) In the case of certainty, when all the probabilities vanish but one which will be equal to unity ($p_1 = 1$; $p_i = 1$ for $i > 1$), the entropy, S_q will be equals to zero ($S_q = 0$). In this case the value of q will be immaterial. This is consistent with the results from Boltzmann-Gibbs formula, Eq. (7).

(5) Generally when two independent systems A and B join, the probabilities in the joined system will have the following relation with the probabilities of each system before joining:

$$p_{ij}^{A+B} = p_i^A \cdot p_j^B$$

Considering this expression in the entropy of the mixture $S(A+B)$, we will get:

$$\frac{S_q(A+B)}{k} = \frac{1 - \sum_{i=1}^W (p_i^A \cdot p_j^B)^q}{q-1} = \frac{S_q(A)}{k} + \frac{S_q(B)}{k} + (1-q)\frac{S_q(A)}{k} \cdot \frac{S_q(B)}{k}. \quad (9)$$

This equation is indicative of the fact that the entropy is non-additive, which is a result of nonextensivity of the systems.

(6) Another important property of the nonextensive systems is the following [36]. Suppose that the set of W possibilities is arbitrarily separated into two subsets having respectively W_L and W_M possibilities ($W_L + W_M = W$). We define

$$p_L = \sum_{i=1}^{W_L} p_i \quad \text{and} \quad p_M = \sum_{i=W_L+1}^W p_i ,$$

hence $p_L + p_M = 1$. In this case we can define conditional probabilities

$$p_{Li} = \frac{p_i}{p_L} \quad \text{and} \quad p_{Mi} = \frac{p_i}{p_M}.$$

Then for these conditional probabilities and using Eq. 8, for entropy of nonextensive systems we can straightforwardly derive the following equation [36,37]

$$S_q(\{p_i\}) = S_q(p_L, p_M) + p_L^q S_q(\{\frac{p_i}{p_L}\}) + p_M^q S_q(\{\frac{p_i}{p_M}\}). \quad (10)$$

This equation plays a central role in the whole generalization of the statistical mechanics of nonextensive systems. Indeed, since the probabilities $\{p_i\}$ are generically numbers between zero and unity then we can conclude that, $\{p_i\}^q > p_i$ for q<1 and $\{p_i\}^q < p_i$ for q>1. Henceforth the entropic index q<1 and q>1 will, respectively, would represent the rare and the frequent events. This simple property lies at the heart of the whole formulation of statistical mechanics of nonextensive systems [36,37].

(7) Another interesting property of the Tsallis entropy, Eq. (8), is the following. The Boltzmann-Gibbs formula for entropy as given by Eq. (7) satisfies the following relation:

$$S = -k\sum_{i=1}^{W} p_i \ln p_i = -k[\frac{d}{d\alpha}\sum_{i=1}^{W} p_i^{\alpha}]_{\alpha=1}. \tag{11}$$

In 1909 Jackson [36,38] introduced the following generalized differential operator for an arbitrary function $f(x)$:

$$D_{\lambda}f(x) \equiv \frac{f(\lambda.x) - f(x)}{\lambda.x - x}. \tag{12}$$

This generalized differential operator satisfies the following limiting condition

$$D_1 \equiv \lim_{\lambda \to 1} D_{\lambda} = \frac{d}{dx}. \tag{13}$$

By considering Eq.s (12) and (13), Abe [39] has concluded the following expression for Tsallis entropy as given by Eq. (8)

$$-k[D_q\sum_{i=1}^{W} p_i^{\alpha}]_{\alpha=1} = k\frac{1-\sum_{i=1}^{W} p_i^q}{q-1} = S_q. \tag{14}$$

This interesting property provides additional insight into the characteristics of the Tsallis generalized entropic form of nonextensive systems, S_q, as given by Eq. (8). Therefore, its connection with Jackson's differential operator appears to be a kind of natural. Indeed, this operator "tests" the function $f(x)$ under dilatation of x, in contrast to the usual derivative, which "tests" $f(x)$ under translation of x.

(8) Finally, let us close the present set of properties reminding that S_q has, with regard to $\{p_i\}$, a definite concavity for all values of q (S_q is always concave for $q>0$ and always convex for $q<0$). In this sense, it contrasts with Rényi's entropy,

$$(\sum_{i=1}^{W} p_i^q)/(1-q) = \{\ln[1+(1-q)S_q]\}/(1-q),$$

which does not have this property for all values of q. Hungarian mathematician Alfréd Rényi constructed the proper entropy for fractal geometries [40].

The entropic index q characterizes the degree of non-extensivity reflected in the above, so called, pseudo-additivity entropy rule [31,34]. The cases $q<1$, $q=1$ and $q>1$ respectively, are named the cases of superadditivity (superextensivity), and subadditivity (subextensivity), respectively. Parameter q is also called the "nonextensive index".

The above expression for entropy, Eq.(8), is applied in many systems including cosmic and nano systems. Eq.s (8) and (9) for entropy can adapt to suit the physical characteristics of many nonextensive systems while preserving the fundamental property of entropy in the Second Law of thermodynamics. Namely, that the entropy production of a system is positive with time in all processes [35]. Although Eq. (8) for entropy reduces to the Boltzmann-Gibbs formula in the case of extensivity as given by Eq. (7), i.e. when the entropy of a system is merely the sum of the entropies of its subsystems. This entropy expression is considered to be much broader than the Boltzmann-Gibbs expression since it describes many nonextensive phenomena including small systems, which are of interest here. As mentioned above the proposed general form of entropy for non-extensive systems, given by Eq. (8) and the entropic index q (intimately related to and determined by the microscopic dynamics), characterizes the degree of nonextensivity of the system.

Microcanonical Ensemble for Nonextensive Systems

The canonical ensemble is characterized by its fixed volume V, fixed number of particles (atoms or molecules) N and fixed absolute temperature T. In other words, V, N and T are the independent variables chosen in the system. The objective here is to calculate the ensemble average values for such mechanical properties of the system as pressure, internal energy, etc. provided the entropy of the system is given by the Tsallis formula, Eq.(8):

Let us now present the connection between the microscopic states of a system and the thermodynamic quantities. We start by assuming that we have a quantum system with discrete energy levels in small systems as we have in large systems. We take as our starting point the Tsallis formula of the entropy as given by Eq. (8),

$$S_q = k(1 - \sum_{i=1}^{W} p_i^q)/(q-1),$$

in which p_i stands for the probability for the system to exist in the quantum state (i). The constant k is still the Boltzmann's constant ($k_B = 1.38054 \times 10^{-23}$ [*Joule/Kelvin*]) which has units of energy divided by temperature. The guiding principle is to maximize the entropy under the constraints as will be presented below.

At first, let us consider maximization of the entropy for an isolated system. If we maximize the entropy, Eq. (8), with respect to the probabilities, we will find that the maximum entropy occurs when all p_i's are the same and they are constant. Let us imagine that there are exactly $\Omega = \Omega(\varepsilon)$ states at equal energy levels of ε (or this many states in a small range of energies centered about ε). The sum of all probabilities must be equal to unity, which is the constraint on the system, so $p_i = 1/\Omega$. This will result in the following maximum entropy expression for the isolated system with respect to its degeneracy Ω:

$$S_q = k \frac{1 - \Omega^{-q+1}}{q-1} \quad \text{and} \quad dS_q = k\Omega^q \, d\Omega \tag{15}$$

This formula connects the microscopic degeneracy to the thermodynamic variable S_q. By joining Eq's (1)-(3) when the system is in equilibrium we arrive at the Gibbs equation,

$$dU = TdS - PdV \tag{16}$$

From Eq. (16) we conclude that

$$\left(\frac{\partial S}{\partial U} \right)_V = \frac{1}{T} \tag{17}$$

and from Eq.s (15) and (17) we derive,

$$\Omega^q \left(\frac{\partial \Omega}{\partial U} \right)_V = \frac{1}{kT} \tag{18}$$

This is a fundamental equation of statistical mechanics of nonextensive systems, which relates the temperature to the change in the degeneracy with respect to energy. Since the degeneracy must increase with energy, the temperature is a positive quantity. Note that the combination

$kT(change\ in)\Omega^q\partial\Omega$ is thermodynamically related to an energy change ∂U. The discussion so far relates to the *microcanonical* partition function: the case where the system is completely isolated from the environment. In the following section, we discuss the case of canonical ensemble for nonextensive systems.

Canonical Ensemble for Nonextensive Systems

Let us now allow the nonextensive system under consideration exchange energy across a boundary wall, which separates it from its surroundings. We maximize the entropy under the constraint that the average energy of the system is a constant. For nonextensive systems, the internal energy constraint is postulated to be

$$U_q = -k\sum_{i=1}^{W} p_i^q \varepsilon_i\ . \tag{19}$$

To perform this optimization with constraints, we employ the method of Lagrange multipliers. We consider the entropy expression modified by two constraints: one for the average energy

$$U_q = -k\sum_{i=1}^{W} p_i^q \varepsilon_i\ ,$$

and one for the normalization of probabilities

$$\sum_{i=1}^{W} p_i = 1\ .$$

This means we minimize the following function S_q',

$$S_q' = k\frac{1-\sum_{i=1}^{W} p_i^q}{q-1} - \beta \sum_{i=1}^{W} p_i^q \varepsilon_i - \theta \sum_{i=1}^{W} p_i\ .$$

It should be noted that β has units of inverse temperature while θ is dimensionless. In order to perform the minimization the following equation must be solved for the probability,

$$dS_q'/dp_i = k\frac{-qp_i^{q-1}}{q-1} - \beta q p_i^{q-1}\varepsilon_i - \theta = 0\ .$$

Then, the following expression for the probability is derived [36]

$$p_i = c[kq/(q-1) + \beta q \varepsilon_i]^{1/(q-1)}.$$

This equation can be written as

$$p_i = [1 - (1-q)\beta \varepsilon_i]^{1/(q-1)} / Z_q,$$

where,

$$Z_q(\beta) = \sum_{j=1}^{W} [1 - (1-q)\beta \varepsilon_i]^{1/(q-1)},$$

is the generalized nonextensive canonical ensemble partition function. With the availability of partition function one can derive the expressions for thermodynamic properties and their relationships with one another. Similar approach as mentioned above can be used to solve for other ensemble theories of statistical mechanics.

Conclusions and Discussion

The historical development of classical thermodynamics of large / macroscopic systems and its applications to a wide range of practical problems took place without any reference at all to the particles comprising the system. It was based only on human experiences with the macroscopic behavior of matter as it relates to conversion of thermal to mechanical energy. This development is entirely rigorous and has the merit of establishing the validity of general thermodynamic principles to all types of matter regardless of its molecular character, but in the thermodynamic limit. However, we need to re-examine classical thermodynamics and its application to small systems before we can apply it to such systems. In this chapter, we have introduced this subject and many more things need to be done to complete it.

It should be pointed out that the problem of predicting and correlating thermodynamic properties of small systems, whether empirically or from molecular properties, is an open and unexplored field of research, while this subject for macroscopic system is in its maturity.

Further research into statistical mechanics for nanoscale systems would lead to possibilities to study the evolution of physical, chemical and biophysical systems on significantly reduced length, time and energy scales. The analytic statistical mechanical modelling and its application to nano systems could be achieved through mathematical modeling employing concepts from statistical mechanics of large systems as well as quantum mechanical many-body theories. It can provide insight into the formation, evolution and properties of nanostructures, self-assemblies and their relation with macroscopic behavior of matter.

Bibliography

[1]. K. E. Drexler, "Nano systems: Molecular Machinery, Manufacturing and Computation", Wiley Pub. Co., New York, NY, (1992).

[2]. M. C. Roco, S. Williams and P. Alivisatos (Editors) "Nanotechnology Research Directions: IWGN Workshop Report - Vision for Nanotechnology R&D in the Next Decade" WTEC, Loyola College in Maryland, Baltimore, MD, September, (1999).

[3]. M. J. Madou, "Fundamentals of Microfabrication: The Science of Miniaturization, Second Edition", 2nd Ed., CRC Press, Boca Raton, FL (2002).

[4]. G. A. Mansoori, "Advances in atomic & molecular nanotechnology" in Nanotechnology, United Nations Tech Monitor, 53, Sep. - Oct., (2002).

[5]. T. L. Hill. "Thermodynamics of Small Small Systems", I, W.A. Benjamin, New York, NY (1963); ibid, II, W.A. Benjamin, New York, NY (1964).

[6]. T. L. Hill, Nano Lett., 1, 5, 273, (2001).

[7]. T. L. Hill, Nano Lett., 1, 3, 111, (2001).

[8]. T. L. Hill, Nano Lett., 1, 3, 159, (2001).

[9]. M. Edalat, G. A. Mansoori, and R. B. Bozar-Jomehri, in "Encyclopedia of Chemical Processing and Design" 61 p. 362, Marcell-Dekker, Inc., New York, NY, (1997); M. Edalat, R. B. Bozar-Jomehri and G. A. Mansoori, Oil & Gas J., 39, Feb. 1, (1993).

[10]. G. A. Mansoori, N. F. Carnahan, K. E. Starling and T. W. Leland, J. Chem. Phys., 54, 4, 1523, (1971).

[11]. H. B. Callen, "Thermodynamics and an Introduction to Thermostatistics", John Wiley & Sons, Inc., New York, NY, (1985).

[12]. P. T. Landsberg, "Thermodynamics", Interscience Pub., New York, NY, (1961).

[13]. P. W. Bridgman, "The Nature of Thermodynamics," Harvard Univ. Press, Cambridge, MA, (1943).

[14]. G. A. Mansoori and F. B. Canfield, Indus. & Eng. Chem., 62, 8, 12, (1970).

[15]. J. M. Haile and G. A. Mansoori (Editors) "Molecular-Based Study of Fluids", Avd. Chem. Series 204, ACS, Wash., D.C., (1983).

[16]. E. Matteoli and G. A. Mansoori (Editors), "Fluctuation Theory of Mixtures" Taylor & Francis, London, UK, (1990).

[17]. G. A. Mansoori, "Organic Nanostructures and Their Phase Transitions" in *Proceed. of the 1st Conf. on Nanotechnology - The next industrial revolution*", **2**, 345, March (2002).

[18]. R. P. Feynman, "*There's plenty of room at the bottom - An invitation to enter a new field of physics*" *Engineering and Science Magazine* of Cal. Inst. of Tech., **23**, 22, (1960).

[19]. L. Onsager and S. Machlup, *Physikalische Verhandlungen*, **3** s. 84, (1952).

[20]. I. Prigogine, "*Introduction to Thermodynamics of Irreversible Processes*", John Wiley &. Sons, New Yor, NY, (1967).

[21]. G.A. Mansoori and A. Suwono, "*Introduction to FTEC*", Proceed. of 4th Int'l Conf. On Fluids and Thermal Energy Conversion, Bali, (2003).

[22]. L. Landau and E. Lifshitz, *Teorý a de la Elasticidad*, Reverte´, Barcelona, Spain, (1969).

[23]. G. M. Wang, E. M. Sevick, E. Mittag, D. J. Searles, and D.J. Evans, *Phys. Rev. Lett.*, **89**, 5, 050601, (2002).

[24]. J. Wilks, "*The Third Law of Thermodynamics*" Oxford Univ. Press, Oxford, UK, (1961).

[25]. D. H. E. Gross "*Microcanonical Thermodynamics*" World Sci. Lect. Notes in Physics, **65**, World Sci. Press, Hackensack, NJ, (2001).

[26]. R. S. Berry, J. Ross, S. A. Rice, S. R. Berry, "*Matter in Equilibrium: Statistical Mechanics and Thermodynamics*" 2nd edition, Oxford Univ. Press, Oxfors, UK, (2001).

[27]. D.A. McQuarrie, "Statistical Mechanics", University Science Books, Herndon, VA, (2000).

[28]. J. Kestin, "*A Course in Statistical Thermodynamics*", Academic Press, New York, NY, (1971).

[29]. D.A. McQuarrie, "*Statistical Thermodynamics*", University Science Books; Reprint edition, Herndon, VA, (1985).

[30]. O. Penrose, "*Foundations of Statistical Mechanics. A Deductive Treatment*", *Intern'l Series of Monog. in Natural Phil.*, **22**, Pergamon Press, New York, NY, (1970).

[31]. C. Tsallis, *J. Stat. Phys.*, **52**, 479, (1988).

[32]. C. G. Chakrabarti and D. E. Kajal, *J. Math. & Math. Sci.*, **23**, 4, 243, (2000).

[33]. A.K. Aringazin and M.I. Mazhitov, *Physica A*, **325**, 409 (2003).

[34]. Q.A. Wang, *Phys. Lett. A*, **300**, 169 (2002); Wang, Q.A. and Le M´ehaut´e, A., *J. Math. Phys*, **43**, 5079 (2002); Wang, Q.A., *Euro. Phys. J. B*, 26, 357 (2002); Wang, Q.A., L. Nivanen, A. Le M´ehaut´e and M. Pezeril, *J. Phys. A*, **35**, 7003 (2002); Wang, Q.A. and Le M´ehaut´e, A., *Chaos, Solitons & Fractals*, **15**, 537, (2003).

[35]. C. Tsallis, *Chaos, Solitons and Fractals* **6**, 539 (1995).

[36]. C. Tsallis, Renio S. Mendes, A.R. Plastino, *Physica A*, **261**, 534, (1998).

[37]. R.J.V. Santos, *J. Math. Phys.* **38** (1997) 4104.

[38]. F. Jackson, *Mess. Math.* **38** (1909) 57.

[39]. S. Abe, *Phys. Lett. A*, **224**, 326, (1997).

[40]. A. Rényi, *"Selected Papers of Alfred Rényi*, Vol.2 (Akad´emia Kiado, Budapest, (1976).

Chapter 4

Monte Carlo Simulation Methods for Nanosystems

"Random Numbers. How are the various decisions made? To start with, the computer must have a source of uniformly distributed psuedo-random numbers. A much used algorithm for generating such numbers is the so-called von Neumann "middle-square digits." Here, an arbitrary n-digit integer is squared, creating a 2n-digit product. A new integer is formed by extracting the middle n-digits from the product. This process is iterated over and over, forming a chain of integers whose properties have been extensively studied. Clearly this chain of numbers repeats after some point. H. Lehmer has suggested a scheme based on the Kronecker-Weyl theorem that generates all possible numbers of n digits before it repeats. (See "Random-Number Generators" for a discussion of various approaches to the generation of random numbers.)
    Once one has an algorithm for generating a uniformly distributed set of random numbers, these numbers must be to transformed into the nonuniform distribution g desired for the property of interest. It can be shown that the function f needed to achieve this transformation is just the inverse of the nonuniform distribution function, that is $f=g^{-1}$." **Nicholas C. Metropolis**

Introduction

The Metropolis Monte Carlo (MC) simulation methods can be used in nanoscience to simulate various complex physical phenomena including property calculations, prediction of phase transitions, self-assembly, thermally averaged structures and charge distributions, just to name a few [1]. There exist a variety of MC simulations, which are used depending on the nano system under consideration and the kind of computational results in mind. They include, but not limited to, Classical MC, Quantum MC and Volumetric MC.

In the Classical MC the classical Boltzmann distribution is used as the starting point to perform various property calculations. Through

Quantum MC simulations one can compute quantum mechanical energies, wave functions and electronic structure using the Schrödinger equation. The Volumetric MC is used to calculate molecular volumes and sample molecular phase-space surfaces.

The basis of MC simulations is random numbers. The most important problem for a successful MC simulation is generating random numbers [1].

Generating Random Numbers

One typically needs to generate random numbers distributed according to some given distribution function. Let us first see how uniformly distributed random numbers can be generated. For the sake of completeness, we will briefly mention how one can generate a set of pseudo random numbers. The prefix "pseudo" is used because the generated numbers are never truly random, since they can be regenerated exactly if the same initial "seed" and computer program are chosen. More importantly, the generated set is always periodic, but one tries to make its period as large as possible to avoid any correlations among the generated numbers [1-3].

In general, however, we recommend the reader to use the compiler's own random, or rather "pseudo random" numbers package, or subroutines written by professionals.

Generating Uniformly Distributed Random Numbers in [0,1)

If there is no stochastic process occurring in the computer chip, it will not be able to generate truly random numbers. All it could produce are "pseudo" random numbers generated according to some specific algorithm. The most widely used algorithm is the "Linear Congruential Random Number Generator" presented by D. J. Lehmer in 1949 [4]. The set of pseudo random numbers r_k may be generated according to the following rule:

$$s_0 = seed; \quad s_k = (g * s_{k-1} + i) \bmod(m), \quad \text{for } k > 0 \quad 0 \le s_k < m$$
$$r_k = s_k/m \tag{1}$$

The numbers r_k thus defined are "pseudo-randomly" distributed in $[0,1)$. The starting point s_0 is called the seed, so that the same set of numbers can be reproduced if one uses the same seed. The symbols g, i, and m are called generator (or multiplier), increment and modulus, respectively. The "mod" function, which C++ implements with %, returns the remainder. For example, 90 *mod* 13, written as 90 % 13 in C++, is 12, the remainder in the division of 90 into 13.

As an example of the application of Eq. (1) let us choose the initial value in the sequence of random numbers, $s_0 = 201$ and $g = 9$, $i = 2$, $m = 13$, then $s_1 = (9*201+2) mod 13 = 1811$ $mod 13 = 4$ and $r_1 = 4/13$, (4 is the remainder in the division of 1811 into 13). After we calculate one random number "4/13", to evaluate the next random number, we substitute s_1 into the expression on the right hand side. Consequently we calculate, $s_2 = (9 * 4+2)$ *mod* $13 = 38$ *mod* $13 = 12$ and the next random number is $r_2=12/13$. This can be continued repeatedly and the result will be the following set of periodic random numbers,

$$\frac{4}{13},\frac{12}{13},\frac{6}{13},\frac{4}{13},\frac{12}{13},\frac{6}{13},\frac{4}{13},\frac{12}{13},\frac{6}{13}\dots\dots.$$

The sequence produced in this way is periodic of *3*, but one can make the period very large by choosing appropriate values for the above parameters. The seed, s_0, should preferably be a large odd number, otherwise, one needs to drop the first few generated points. The choice of $m = 2^E$ simplifies the mod operation: in the binary representation of a given number, one only needs to keep the modulus m with most digits. Exponent E should be chosen as large as possible: $E>35$ would be a good choice to make the period as large as possible. Usually, it is taken as the number of bits in the largest positive integer. A large value for g will reduce serial correlations. The number i must be prime to m; $g-1$ should be multiple of every prime factor of m. In this case the period of the sequence would be 2^E. For more details see [1,3].

Generating Random Numbers in [a,b) According to a Given Distribution Function P(x)

Importance Sampling: Also known as "biased sampling" can be used for generating random numbers according to a given distribution function *P(x)*. This is because the efficiency of random sampling methods can be appreciably enhanced by a suitable choice of sample points. The procedure described is called "importance sampling".

The purpose behind importance sampling is to concentrate on the sample points where the value of the function is appreciable and avoid taking sample points where the value of the function is quite negligible. It is necessary to allow for this bias in the sample points by weighting the sample values appropriately. The general approach to importance sampling consists of two steps:

(1) The first step is to carry out a preliminary evaluation of the function. The region of interest is divided into fairly large cells, the value of the function is determined at the center of each cell, and this value and the cumulative sum are recorded.

(2) The next step is to take as many sample points as is desired, but to adapt the sampling procedure so that the number of sample points in each cell is proportional to the function value at the center of that cell, as determined in step (1). However, each sample is given a weight inversely proportional to the central value of its cell.

There are a few ways to generate random numbers through importance sampling. We will mention here four such methods.

(i) For the sake of simplicity, we can mention one simple algorithm, which can easily be implemented and used in one dimension. For a given interval [a,b) and desired number of random points N, one will first divide the interval into $\sqrt{N}$ pieces (this number is somehow arbitrary), and then generate a uniform set of points in each piece, the number of generated points being

$$N P(x_i) \bigg/ \sum_i P(x_i),$$

where x_i is a point in the middle of each piece. Clearly, the generated points are distributed according to P and they are almost N in number. It must be noted, however, that the generated points are strongly correlated in position and need a reshuffling before being used.

(ii) It is possible to analytically map the uniform random numbers in $[0,1)$ to $[a,b)$ if the distribution function $P(x)$ is simple enough. If r is uniform in $[0,1)$, then we have

$$dr = P(x)dx \quad \Rightarrow \quad r = \int_a^x P(t)dt = g(x,a); \qquad g(b,a) = 1 \quad (2)$$

This relation needs to be inverted if possible to obtain x as a function of r. If r is uniform in $[0,1)$, then x will be distributed in $[a,b)$ according to P.

(iii) For an arbitrarily complicated function P, a third way is to use the well-known **Metropolis algorithm** [5]. In this algorithm, one can start from an arbitrary point in $[a,b)$, say $x_0=(a+b)/2$, and add a random number to it: $x_1=x_0+d(2r-1)$ where r is the random number in $[0,1)$ and d is the magnitude of the step. If $P(x_1)/P(x_0)>1$, then the move is accepted; otherwise the ratio is compared to a random number in $[0,1)$. If the ratio is greater, then again, the move is accepted, otherwise it is rejected and the old point x_0 is kept, and one starts another trial from this point. There is also the possibility that the generated points go out of $[a,b)$ range in which case, they will be rejected. Compared to the first method, this is by far less efficient because there will be many rejections. The efficiency of this method depends strongly on the choice of step size d. With a small d most trials will be accepted, but a good sampling will require many trials. Whereas if d is large, one can sample the $[a,b)$ interval quicker, but there might be many rejections.

(iv) A more efficient method due to Von Neumann proposed in 1951 consists of considering a function f {for x in $[a,b)$; $f(x) > P(x)$} for which one can use method (ii) presented above, and generate points distributed according to f (see Figure 1). The simplest choice would be to take f=constant; but larger than the maximum of P. Once such point x is generated in [a,b), one compares a random number r in [0,1) to the ratio $P(x)/f(x)$. If r is less than the ratio, the number x is accepted, otherwise it is rejected and one should repeat the process with another point x distributed according to f. For constant f, the probability of accepting x is clearly proportional to $P(x)$.

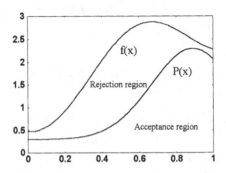

Figure 1. Von Neumann algorithm involves selecting 2 random numbers in [0,1], the first one, distributed according to f, is the x-coordinate; and the second one, uniform, the y-coordinate between 0 and f(x). If it falls below P(x) the point is accepted [1].

Monte Carlo Integration Method

Standard deterministic methods of numerical integration of a function are well known (see for example [6]). The simplest method is the trapezoidal rule, which adds up the function values at equally separated points and multiplies the result by the width of the intervals. A more accurate method is the Simpson's rule, which, instead of approximating the function by linear segments on each interval, replaces it by a quadratic function. Even a more sophisticated method is the Gaussian quadrature, where the integral is the weighted sum of the function evaluated at certain special points. These methods, even though quite accurate for one-dimensional (single) integrals, soon become very prohibitive in multi-dimensional (multiple) integrals as the required number of function evaluations grow quite rapidly as N^d. N is the number of function evaluations required for a one-dimensional integral and d is the number of dimensions in a multi-dimensional integral. In this situation, Monte Carlo integration using random numbers can become very handy.

Regardless of the dimension, one can approximate the integral by a sum in which the function is evaluated at random points:

$$I = \int_a^b f(x)\,dx \approx S = \frac{b-a}{N}\sum_{i=1}^{N} f(x_i). \tag{3}$$

The points x_i are randomly and uniformly distributed on the [a,b) interval. As the number of points N grows, the central limit theorem (CLT) guarantees the convergence of the sum to the real value of the integral. The error being of the order of $1/\sqrt{N}$. More precisely, the CLT states that if we repeat this process of generating N random numbers and computing the above sum S, we generate a set of random variables S. These variables have a Gaussian distribution of mean I (the real value of the integral) and variance $\sigma = \bar{v}/\sqrt{N}$ where $\bar{v}$ is the mean deviation of $(b-a)f(x)$ from the mean value I. The value of $\bar{v}$ can be estimated, however, by considering the deviation from the mean in a given sample (the usual standard deviation). Of course $\bar{v}$ defined as above is sample-dependent; one therefore can take its average over all the taken samples. The CLT implies that as the number of samples is increased, the mean obtained from the samples converges to the real mean $I \sim 1/\sqrt{m}$, where m is the number of samples, and the uncertainty on the accuracy of the result is reduced. The error may be estimated from the CLT:

$$I = \int_a^b f(x)\,dx \approx S = \frac{b-a}{N}\sum_{i=1}^{N} f(x_i) \pm (b-a)\sqrt{\frac{\langle f^2 \rangle - \langle f \rangle^2}{N}}, \tag{4}$$

where the numerator under the square root may be estimated from the taken sample. One can thus reduce the error by either increasing N or reducing the mean standard deviations of f.

The above integration method, although convergent, might not be very efficient if the function has large variations in the interval. Consider for instance $f(x) = e^{10x}$ in the interval [0,1); clearly a large number of points and function evaluations will be "wasted" near $x=0$ where the function is much smaller (by more than 4 orders of magnitude) compared to points near $x=1$. It would be better to sample the interval in a non-uniform fashion so as to have more points near $x=1$ and fewer points near $x=0$. This is what importance sampling can do. We choose a non-uniform distribution function $P(x)$ according to which the points x_i will be sampled. The integral can also be written in the following form:

$$I = \int_a^b f(x)\,dx = \int_a^b P(x)\,[f(x)/P(x)]\,dx \approx \frac{b-a}{N}\sum_{i\in P}^N f(x_i)/P(x_i). \quad (5)$$

Notice that now the function f/P is integrated on points distributed according to P. The function P is of course analytically known and normalized to unity. There would be less fluctuations in the integrand f/P if the distribution function P, which we could choose at will, looks like the original function f.

At any rate, the central idea in importance sampling is to choose P so that it looks like f and therefore f/P does not fluctuate much and the variance of the result is small. In other words, the estimated value of the integral is accurate, even with a relatively small number of points. Table 1 shows a comparison of the estimation of the integral using uniform sampling and importance sampling of the function $f(x)=e^{10x}$ with a quartic distribution function [1] in $[0,1)$.

Table 1: Results [1] of the integral $I = \int_0^1 e^{10x}\,dx = \dfrac{e^{10}-1}{10} \approx 2202.5466$

using uniform and importance sampling with $P(x) = 5x^4 + 0.0001$.

N	Uniform	Importance	Exact
100	2243.5107	2181.7468	2202.5466
1000	2295.6793	2208.6427	2202.5466
10000	2163.6349	2207.1890	2202.5466
100000	2175.3975	2203.7892	2202.5466
1000000	2211.7200	2202.8075	2202.5466
10000000	2203.8063	2202.5431	2202.5466

Note that we preferably want the distribution function P to be non zero in the regions where the function f is non zero, otherwise the ratio f / P leads to infinity if a random point is generated in that region. This is called undersampling and may cause large jumps in f / P in the regions where $P(x)$ is too small. This is the reason why a small number was added to $P(x)$ to avoid undersampling near $x=0$.

The error can be estimated as the following: The average deviation of the function from its mean value is about 5000 as can be seen in the plot of e^{10x} in $[0,1)$. However, a better estimate for the average deviation can be computed from the deviations from the numerically calculated mean (or integral) so $v \approx 5000$. Therefore, the standard deviation is of the order of $5000/\sqrt{N}$ for uniform sampling (from the CLT). As for importance sampling, one must evaluate the mean standard deviations of the function f / P in $[0,1)$, and then multiply it by $1/\sqrt{N}$. We can conclude from the above numerical data that the agreement of the last line to within 6 digits is not expected, and therefore, fortuitous. The following is the statement of CLT in mathematical terms:

$$I = \int_a^b P(x)f(x)/P(x)\,dx \approx S$$

$$= \frac{b-a}{N}\sum_{i \in P}^{N} \frac{f(x_i)}{P(x_i)} \pm (b-a)\sqrt{\frac{\left\langle (f/P)^2 \right\rangle - \left\langle f/P \right\rangle^2}{N}}. \tag{6}$$

From this formula, it is clear that the error is much reduced if the function f / P has weaker fluctuations than f itself. Note that since functions f change sign, and since P has to be always positive, the ratio f / P will always have sign fluctuations. Therefore, variance will never go to zero even if N goes to infinity. This problem can be overcome only if the roots of f are known and the integral can be split to sums of integrals over regions where the sign of f does not change.

Applications to Nanosystems Composed of a Few Particles

Variational Monte Carlo Computation: In this method the Monte Carlo integration technique is used to evaluate multi-dimensional integrals which appear in the evaluation of the ground state energy, E_o, of many particle systems such as few electrons in an atom, a molecule or a quantum dot [7,8]:

$$E_0 = \frac{\int \Psi^* H \Psi \, dR}{\int \Psi^* \Psi \, dR}. \tag{7}$$

In this equation, H is the Hamiltonian and

$$\Psi(r_1, r_2, ..., r_N) = \sum_{i=0}^{\infty} c_i \psi_i,$$

is the trial wavefunction, where the expansion coefficients,

$$\sum_{i=0}^{\infty} |c_i|^2 = 1,$$

are normalized. The trial wavefunction contains a number of variational parameters. The integral (7) may be rewritten in an importance sampled form in terms of the probability density distribution function $|\Psi|^2$ in the following form:

$$E_0 = \frac{\int |\Psi|^2 \frac{H\Psi}{\Psi} \, dR}{\int |\Psi|^2 \, dR}.$$

The energy is then minimized as a function of the latter. This is the basis for the most accurate calculations of the ground state properties of a many-particle quantum system.

Importance sampling can also be used to compute the partition function and average of thermodynamic quantities of a few interacting particles in a finite box [8]:

$$E = \frac{3}{2} NkT + \langle \phi \rangle = \frac{3}{2} NkT$$

$$+ \int dr_1 ... dr_N \quad \phi(r_1, ..., r_N) \, e^{-\beta V(r_1, ..., r_N)} \Big/ \int dr_1 ... dr_N \, e^{-\beta \phi(r_1, ..., r_N)} . \quad (8)$$

The typical way importance sampling is done in such calculations is to generate a series of random points distributed according to $\left| \Psi^2 \right|$ in the quantum case and to $e^{-\beta \phi(r_1, ..., r_N)}$ in the classical case. In the former, one then computes an average of the "local energy"

$$E(R) = \hat{H}\Psi(R) / \Psi(R); \quad R = (r_1, r_2, ..., r_N),$$

and in the latter, just the average of the potential energy $\phi(R)$. Notice that this is just a choice, and one could sample the points R in the 3-dimensional space according to any other distribution function. As said before, the idea is to sum a function $\{E(R) \text{ or} \phi(R)\}$, which has as little fluctuations as possible. Therefore, if the above functions still have large fluctuations, one could use another importance sampling provided that the distribution function is analytically integrable so that the weight is properly normalized.

One way such averages maybe done is, for instance, to assume a simple function $W(R)$ which "looks like" $\phi(R)$ at least near its minima, and whose exponential is integrable (a quadratic function, the harmonic approximation of $\phi(R)$, comes naturally to mind), use sampling points distributed according to $e^{-\beta W(R)}$, and sample the function,

$$\phi(R) e^{-\beta [\phi(R) - W(R)]} .$$

However, this same function will sample the numerator and denominator simultaneously,

$$Z = \int dR \, e^{-\beta \phi(R)} = \int dR \, e^{-\beta [\phi(R) - W(R)]} \, e^{-\beta W(R)} , \quad (9)$$

so that both are computed simultaneously. The advantage of this method is that it will also compute the partition function, which might be of use in some other calculations. In choosing W, one must be careful, however, not to under sample: the "harmonic potential", i.e. W must not become much larger than the true potential in the regions away from the minimum. Note that instead of the Boltzmann distribution, used for macroscopic systems, one can choose any other distribution function such as the one by Tsallis, introduced in Chapter 3, to sample the phase space.

The real difficulty in computing integrals accurately, besides the problem with appropriate choice of the weighting function P, is to generate uncorrelated random numbers that are distributed according to P. This is the subject of the following paragraphs in which we will make use of the famous Metropolis algorithm (already mentioned as method (iii) for importance sampling) along with the Markov process.

Equilibrium Statistical Mechanics and Monte Carlo Method

The most widely use of MC method has been in solving problems in equilibrium statistical mechanics (SM) calculations [7-11]. Both fields MC and SM naturally deal with statistics and averaging such as the one mentioned above under "Variational Monte Carlo Computation". In equilibrium statistical mechanics, one is interested in generating an ensemble of equilibrium states so that averages of physical properties can be numerically computed.

The Markov Process

The **Markov process** is a stochastic process by which one goes from a state i at time t to a state j at time $t + \Delta t$. Its future probabilities are determined by its most recent values and not by how it arrived in the present state.

The walk towards equilibrium from an arbitrary initial state, is done by using a **Markov Process**. The same process is then used to generate the ensemble of equilibrium states. To avoid correlation between the generated states in the ensemble, one must only take very few of the generated states in order to perform averages; because in a "continuous" walk, points (states) following one another are correlated.

What we mean by "walk" in this Markovian dynamics, is going from one state of the system to another by some stochastic process. If this walk is done correctly, meaning any state is within reach (**Ergodicity**), and provided the walk is continued long enough, then after reaching some **relaxation time,** one reaches the equilibrium state and can perform the needed statistics.

Ergodicity means that all states can be reached, in principle, if we wait long enough. This is required to guarantee that we will not miss the ground state or get stuck in some local minima and not be able to make the correct statistics at equilibrium.

The relaxation time is the time required for a system to recover a specified condition after a disturbance. In stochastic processes, it is a measure of the rate at which a disequilibrium distribution decays toward an equilibrium distribution. It should be mentioned that there is no simple way of finding the relaxation time in advance. The easiest way is to let the system evolve once, then from the time series of different physical quantities, estimate the relaxation time.

The distribution function of the system out of equilibrium usually satisfies a time evolution equation known as the **master equation**. It describes how the system jumps from one state to another:

$$\frac{dp_i}{dt} = \sum_j (p_j \theta_{j \to i} - p_i \theta_{i \to j}). \tag{10}$$

Here $\theta_{i \to j}$ is the transition rate (or probability if properly normalized) to go from the state i at time t to the state j at time $t + \Delta t$. Assuming the states (which are infinite in number), to be discrete, one can think of θ as a matrix which we also call the **transition matrix**. Notation $\theta_{i \to j}$ stands for the matrix element in line j and column i, and $p_i(t)$ is the probability of finding the system in the state i at time t. The transition matrix θ follows the sum rule (in units of the time step Δt),

$$\sum_{j} \theta_{i \to j} = 1. \tag{11}$$

This simply means that the sum of the transition probabilities from a given state i to all other states is unity.

The matrix θ might be known for some systems if the physical processes inducing the transitions are known. For generating an ensemble of states at equilibrium, one needs to design this matrix. If θ is known, we can simulate the time evolution of the distribution function. This will be the subject of non-equilibrium MC simulations, which will be discussed later.

Accordingly, for a Markov process the corresponding transition matrix must have the following properties: The transition rate should not depend on the history of the system, and at any time the transition probability $\theta_{i \to j}$ must always be the same regardless of what the states of the system were at prior time steps; i.e. it only depends on the states i and j. In order to reach the correct equilibrium state of the system, we require the transition matrix, which is modeling the Markov process to be **Ergodic** and satisfy the requirements of **Detailed Balance**.

Detailed balance ensures that the final stationary state is the thermodynamic equilibrium state. The stationary state condition is when $dp_i / dt = 0$ meaning we can have transitions in and out of a state, but they add up to zero and there is no net change, i.e.

$$\frac{dp_i}{dt} = 0 \Rightarrow \sum_{j} p_j^{eq} \theta_{j \to i} = \sum_{j} p_i^{eq} \theta_{i \to j}. \tag{12}$$

Then to satisfy the detailed balance condition, it is sufficient for the transition matrix to satisfy the following condition

$$P_i^{eq} \theta_{i \to j} = P_j^{eq} \theta_{j \to i} \quad \text{(For all } i \text{ and } j \text{ states)}. \tag{13}$$

To make sure detailed balance is satisfied and no cyclic motion appears in the "equilibrium" state, we can impose Eq. (13) which satisfies the microscopic reversibility of θ. While Eq. (14) satisfies the detailed balance, Eq. (12), it is not a necessary condition for detailed balance. In other words, microscopic reversibility is more constraining than detailed balance. Equation (13) simply indicates the currents from i to j cancel the currents from j to i, which is a sufficient condition for being in steady state.

Now let us discuss the form of the transition matrix that we would like to adopt. If the transition matrix can be estimated, or approximated in some way, then it would be possible to obtain the equilibrium distribution. We can also get an estimate of the relaxation time, which contains information about the dynamics of the interactions occurring in the system. It is also possible to obtain the relaxation time from the transition matrix.

Usually we know the equilibrium distribution we want to reach. In most instances, we are dealing with the Boltzmann distribution where $p_i^{eq} = e^{-\beta E_i} / Z$, $\beta = 1/kT$ and Z is the partition function,

$$Z = \sum_i e^{-\beta E_i}.$$

One can therefore deduce from the detailed balance condition the ratio

$$\theta_{i \to j} \big/ \theta_{j \to i} = p_j^{eq} / p_i^{eq} = e^{-\beta(E_j - E_i)}. \tag{14}$$

The last point, but not the least, about sampling using a Markov process is the correlation in the time series. When performing ensemble averages, data separated by a time segment larger than the correlation time should be used.

The correlation time in a Markov process is the time over which a certain particle distribution pattern is correlated. In a Markov process undergoing fluctuations, individual fluctuations appear and disappear over a characteristic correlation time. Mathematical methods exist in the form of the autocorrelation and cross correlation functions, which are used to extract information about the correlation time. The correlation time is when the autocorrelation function of a given property decays to zero.

Choice of the Transition Function

There are many possible choices for the transition matrix with the given constraints of detailed balance and ergodicity and the sum rule as given by Eq. (11). If one thinks of θ as a matrix for which the upper left triangular part is known, detailed balance yields the lower right triangular part. The sum rule guarantees the unity of the sum of the elements in a column, and thus defines the diagonal elements. Ergodicity enforces that not all elements (other than the diagonal terms) in a row/column are zero. Otherwise, one will never transit to the state corresponding to that row/column. Therefore, once the upper triangular part is carefully defined, one can obtain all other transition probabilities.

The most widely used transition matrix is the one proposed by Metropolis, *et al* [5]. In this work, for the first time in 1953 these authors used the MC method to numerically compute the equation of state of a hard–disk system (in two dimensions). In the Metropolis algorithm, the transition probabilities are given by:

$$\theta_{i \to j} = Min(1, p_j^{eq} / p_i^{eq}).$$ (15)

This means the move from state i to state j is accepted if the latter is more probable, otherwise it will be accepted with the probability P_j^{eq} / P_i^{eq}. In practice, the way this is done is to pick a random number, r, in [0,1) range; the move is rejected if $r > p_j / p_i$, otherwise it will be accepted. In case of rejection, one starts from the state i and tries another move, and when doing the statistics, the state i must be counted as many times as the moves were rejected. It could be easily verified that equation (15) satisfies the detailed balance condition.

Although Metropolis sampling of the phase space works for almost any problem, the fact that there is rejection in the moves makes the algorithm rather slow, and for any specific problem, one can usually devise better and more adapted algorithms. Some of these more advanced methods will be discussed later in this Chapter. Mc Millan [12] to treat a quantum system by the variational method used the Monte Carlo method combined with the Metropolis algorithm for the first time in 1964. Below, is an illustration of the results obtained for a nano system by this method [1].

Example

One application of the Monte Carlo method is to sample the phase space and produce an ensemble of equilibrium states at a given temperature *T*. As an illustration, in Figure 2 the graphs of the energy fluctuations and distribution function obtained for a cluster composed of thirteen Argon atoms interacting via the Lennard-Jones pair intermolecular potential energy function are reported [1].

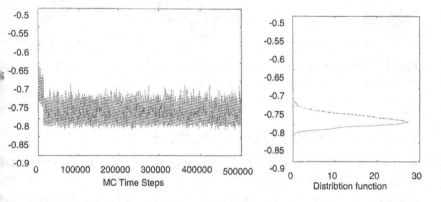

Figure 2. The graphs of the energy fluctuations and distribution function obtained for a 13-atom cluster of Argon atoms interacting via the Lennard-Jones potential [1].

After some relaxation time, the ground state configuration, which is an icosahedron, is obtained. Then averages can be performed, after the ground state structure is reached completing 20,000 steps as shown in

Figure 2, in order to obtain the average energy *E(T)* and the heat capacity. Heat capacity can be deduced either from the derivative of *E(T)* with respect to temperature or, even better, from its fluctuations, i.e. [1]:

$$C = \frac{\partial \langle E \rangle}{\partial T} = \frac{\langle E^2 \rangle - \langle E \rangle^2}{k_B T^2}; \text{ using the definition } \langle E \rangle = -\frac{\partial \text{Log} Z}{\partial \beta}.$$

In Figure 3, three configurations of the cluster at the initial, middle and final steps are shown.

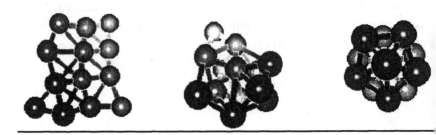

Figure 3. Three configurations of a 13-atom Lennard-Jones cluster at the initial, middle and final steps of an MC run, from left to right, respectively [1].

Acceptance Ratios and Choice of the Moves

In a Markov stochastic process one generates certain trial moves and accepts them with a certain probability. The choice of trial moves depends on the nature of the system that is under consideration. To obtain maximum efficiency the correlation between successive configurations is made as small as possible.

For example, when dealing with discrete degrees of freedom such as in the Ising model, a move could consist of a single spin flip at a time. This could turn out to be very slow if one is near the Curie temperature where fluctuations are large and all such moves will almost be accepted. Flipping many spins at the same time could also result in many rejections especially if the temperature is low. So, depending on the state variables (temperature, etc.), different choices need to be made for making efficient moves in the phase space. A useful criterion in this case, is the **Acceptance Ratio** [7,13]. It provides the percentage of the successful tried moves was during a given number of trials. A good rule of thumb for Metropolis algorithm is that, moves should be chosen such that the acceptance ratio is kept between 20 to 50%.

One has the choice on making the move irrespective of its probability. This freedom in choice can allow one to make moves with high acceptance probabilities, and thus have an algorithm, which can be more efficient than the simple "blind" Metropolis algorithm. For instance, when sampling the canonical ensemble, moving to points with high energy has a good chance of being rejected. One could instead, move to points with "not so high" energy, with a higher chance of being accepted. This presumes knowledge of the local energy landscape. If one has such information, it must be used in order to have an efficient sampling. Therefore, although Metropolis algorithm works almost for any problem, one can possibly find a more suitable moving algorithm for a specific problem, which would be more efficient in terms of sampling the largest part of the phase space in the shorter time and having a smaller variance.

One such example is the Swendsen-Wang algorithm [13,14], which is used near the Curie temperature in a ferromagnetic system.. Indeed, at the transition temperature, fluctuations in the magnetization are large and the correlation length diverges. Therefore, a single spin flip is considered as a small move and is usually accepted. The idea is to use multiple spin flips as in a cluster flip. However, one cannot consider solid clusters of the same spin since they would always keep their shape. Therefore one needs to connect parallel spins with some probability p. Choosing the appropriate probability p, $0 < p < 1$, is the key to this fast algorithm. It can be shown that the optimal value of p satisfies $1 - p = e^{-2\beta J}$. Clusters constructed with this value of p will always be flipped (large moves with acceptance ratio=1) and one obtains a fast and good sampling of the canonical ensemble even near the Curie temperature.

In conclusion, the best choice algorithm for a given problem is generally problem-specific. In another words, one cannot design an algorithm, which works well for all problems. That is why development of simulation algorithms is still an active field of research.

Other Tricks to Improve the Simulation Speed

In the case of simulation of statistical properties of discrete systems, such as an Ising lattice, the walk towards equilibrium is usually done with single spin flips. In this case, it is not necessary to recalculate the total energy at each step, but just its change due to a single flip, i.e.

$$\Delta E(i) = -2\sum_{j} J_{ij}S_j = -2\sum_{j\,near\,i} J_{ij}S_j . \tag{16}$$

This is a finite sum since the exchange integral J is usually short-ranged. All possible neighboring configurations can be considered and all possible energy changes and transition probabilities $e^{-\beta\Delta E}$ can be computed and tabulated once and for all before the MC run, so that in case of need they will only be looked up and not computed every time.

Another way to speed up the MC runs is to use the **Heat Bath Algorithm** [9]. For a spin chosen at random, one can compute the local probability distribution directly; meaning the energies of all possible states of the spin and therefore the Boltzmann probabilities of being in these states are computed. By generating a single random number r, one can decide in which "local equilibrium state" the spin will be. If there are m possible states $1 \leq j \leq m$, and

$$\sum_{l=1}^{j-1} p_l \leq r < \sum_{l=1}^{j} p_l; \qquad p_l = e^{-\beta E_l} / \sum_{i=1}^{m} e^{-\beta E_i} \tag{17}$$

Then the state j is chosen as the next equilibrium state. This will be done for all spins many times until the system relaxes to equilibrium after which a relevant statistics may be done. The advantage of this method is that the local equilibrium is accessible in one-step. The disadvantage of this method is that updates are still local, and the probability distribution may be difficult or time consuming to compute.

There are also what are called "**faster than the clock algorithms**". Instead of trying flips at every time step and computing transition probabilities (Boltzmann factors), one can compute how many steps it takes before a flip occurs. Imagine the simple case where the transition rate is a constant: the probability of transition between time t and $t + \Delta t$ is $\Delta t / \tau$ as in the problem of radioactive decay. Therefore the probability of no transition from 0 to t, and transition between t and $t + \Delta t$ is $p(N\Delta t) = (1 - \Delta t / \tau)^N \Delta t / \tau$ so that the total probability is

$$\sum_{N=0}^{\infty} p(N\Delta t) = \sum_{N=0}^{\infty} (1 - \Delta t / \tau)^N \Delta t / \tau = 1. \tag{18}$$

To find out when the transition occurs, one needs to generate a random number r and find N for which $p(N\Delta t) > r \geq p((N+1)\Delta t)$. This implies

$$N \approx Log(r\tau / \Delta t) / Log(1 - \Delta t / \tau) \Rightarrow t / \tau = Log(\Delta t / \tau) - Log\, r. \tag{19}$$

This is called faster than the clock algorithm because one can find out the time of transition and flip the particle before the "natural time" that it needs! Another advantage of this algorithm is that it has no rejection. Furthermore, it's output (flipped moves and their times) can not be distinguished from a Metropolis algorithm. These kinds of algorithms are also called **Dynamical Monte Carlo methods** [13,15].

Application of Monte Carlo to Nonequilibrium Problems

The Monte Carlo method can be used for simulating the time evolution of a stochastic system in which the transition rates are known. Here, in contrast to the previous section, which dealt with equilibrium where we could choose the transition rates at will (if detailed balance and ergodicity were preserved), the transition rates cannot be arbitrary. Their choice will represent a real, non-equilibrium physical phenomenon dictating the dynamics of the system and must be given in advance.

In physical processes, the transition matrix is not arbitrary, but can be derived or approximated somehow. For example, let us consider the process of simulating the diffusion of an atom on a surface. Since the atom can move to the left, right, up or down (simple model), each with a probability (which depends on the barriers that the atom needs to overcome) divided by $k_B T$; one can define four transition matrix elements for these four processes. They would be the four Boltzmann factors, which are known if the four diffusion barriers are known. This is what is meant by "transition matrix of a real process known in advance".

Typical problems that are solved using this method are issues of various transport property calculations. Examples are carrier transport in electronic devices dealing with solving the Boltzmann transport equation [1,16], energy, mass and momentum transports, aggregation and growth [17,18].

To the master equation, which represents the time evolution of a distribution function, one can associate an equation of motion for particles subject to external forces, and a stochastic term (in case of contact with a heat bath): this is the Langevin equation.

The Langevin Equation

The Langevin equation is used to describe the Brownian motion of a particle in a fluid. Instead of looking at the variation of velocity over time, as in the Wiener equation (A simple mathematical representation of Brownian motion, assuming the current velocity of a fluid particle fluctuates randomly), the Langevin equation deals with the temporal change in acceleration due to a stochastic force:

$$m\frac{d^2x}{dt^2} = -\frac{m}{\tau}\frac{dx}{dt} + f(t) \tag{20}$$

where m is mass of the particle, f is the stochastic fluctuating force and m/τ is the friction coefficient derived from the Stoke's law derived by solving the Navier-Stokes equation for low Reynold's number.

If the particles follow this equation, then their distribution function obeys the master equation. Therefore, instead of solving the master equation to find the time evolution of the distribution function, one can consider a set of particles and move them according to the corresponding Langevin equation. One can then record the quantities of interest, such as total energy, density, magnetization, current, etc, as a function of time. These quantities are needed since ultimately, one wants to compute statistical averages of theses physical observables. For simplicity, in the present discussion, we will consider non-interacting particles. In this case, the many-particle distribution function can be factored into the product of single-particle distribution functions. One can then only consider the Langevin equation for a single particle, simulate the time evolution of the particle and record averages. For interacting systems, it is also possible to use mean-field theory, and do the same kind of factorization. Many particles are moved in the external plus random field, to which one should also add the field due to other particles.

A simple example is the study of transport using the Boltzmann equation. This equation can be solved in some limits, or within some approximations such as the relaxation time approximation. The obtained solutions, though not very accurate, can give insight into the physics of the problem. In more complex cases where many scattering processes are involved and an analytical solution can not be found, accurate solutions can be obtained using the dynamical (also called kinetic) MC method to numerically integrate the master equation (again by considering a population of particles). We will give below the general algorithm for finding the time evolution of the distribution function in a very general and abstract way, though in practice, one might use a Langevin type of equation. This will give a general idea of how such a simulation should be performed. For now, we will not be concerned about the efficiency of such an algorithm.

Consider the master equation (10) with known transition rates,

$$\frac{dp_i}{dt} = \sum_j (p_j \theta_{j \to i} - p_i \theta_{i \to j}) \; ; \text{(where } i = 1,...,s \text{)},$$

each represent a different (here discretized) state of the system. For a non-interacting system, s is just the number of one-particle states. If one is dealing with a particle in a box, the box can be discretized into a lattice, and then s will be the number of lattice sites times the number of possible momenta, which would depend on a chosen momentum cutoff. It is obvious that soon this number becomes very large. For an N-particle system, the number of possible states is $S = s^N$, and the state index i is going from 1 to S. The state of the system can be thought of as one filled box among the S available boxes. The initial box i is chosen at random; then the particle is moved from this box to another box according to the transition rules defined by $\theta_{i \to j}$. For a non-interacting system, the transition matrix is fixed and can be calculated at the beginning of the run, once and for all. Otherwise, in an interacting system, it needs to be updated at each time step. Once the transition matrix is known, and the present state is known, the state at the next time step is chosen using a random number r such that $0<r<1$: A new state j among the possible S states will be chosen if:

$$\sum_{l=1}^{j-1} \theta_{i->l} \le r < \sum_{l=1}^{j} \theta_{i->l} \; \text{(Since } \sum_{l=1}^{S} \theta_{i->l} = 1 \text{)}. \tag{21}$$

If the transition matrix ($S \times S$) is known and calculated at the beginning only once, then the state at the next time step is simply obtained by choosing a random number r, such that $0<r<1$, and determining r corresponds to transition $i \rightarrow j$. This is not true for an interacting system in which the transition rates might depend on the present state of the system, in which case this matrix must be updated at every time step. This process can be iterated on every, j^{th}, column of the transition matrix, until the equilibrium is reached and statistics can be performed. The process of updating the state has no rejection; each moving step does not correspond to any physical time step, as some of the chosen rates are large and some are small; furthermore, physical parameters of the system, such as temperature, are included in the transition rates.

For a real system, many of the transitions are quite improbable or simply impossible because there are a huge number of states, and therefore the θ matrix has a majority of zero terms. One can only transit from a state j to very few available states i; therefore the transition matrix θ is nonzero only for the available states, and equal to zero the rest of the time.

This algorithm, although very general and straightforward, is impractical in most cases because of the large number of states the system possess. It can be refined, however, in many cases. In most practical cases, a set of particles are assumed and moved according to the Langevin, or some other stochastic, equation corresponding to the Master equation (10). From their trajectories, after the relaxation time has elapsed, the steady state distribution function and other averages can be computed. The short time equilibration process may also be observed and turn out to be useful in such simulations. This method, also called Kinetic Monte Carlo, is used as well to investigate diffusion and growth processes, which take too long a time to be simulated by Molecular Dynamics methods.

Interacting Systems

Consider a population of N interacting particles subject to external fields and undergoing scattering and diffusion processes of known rates. External and mean fields appear in the left side of the Master equation where a time derivative is present, such as in the Boltzmann equation. Scattering and diffusion processes appear in the right side of the Master equation where transition rates are present. Assume that there are $i=1,2,\ldots,p$ types of different processes each with a given rate w_i.

For a non-interacting system, one can consider only one particle, and decide by using a random number the process it will undergo: if $0 < r < 1$ is random, and

$$\frac{1}{z}\sum_{i=0}^{j-1} w_i \leq r < \frac{1}{z}\sum_{i=0}^{j} w_i; \quad \text{where} \quad z = \sum_{i=1}^{p} w_i + w_0,$$

then the process w_j is chosen, w_0 is a fictitious rate called the "self-scattering" rate to allow the computation of "free flight-" or "waiting-" time of the particle. If the process w_0 is selected, then there is no scattering at all and the particle continues its "free flight". Then the particle is scattered according to this process into some other state, which might also require a random number to be selected. Next, one has to find out how long it takes before the particle undergoes the next scattering. This time depends clearly on the present scattering rates. As explained in a previous example, in the section on "Other Tricks to Improve the Simulation Speed", the waiting time is calculated according to

$$t = -\ln r / \sum_{l=0}^{p} w_i,$$

where again r is random in $[0,1)$. This only causes to advance the "physical clock" by t, whereas the "simulation clock" is only incremented by one. Again a new scattering process must be chosen at random, and the above-mentioned stages must be repeated.

This needs to be done for a long time on a single particle if the system is non-interacting or on all particles if interactions are present. Then ensemble averages are replaced by time averages. If large amounts of memory and a small number of states are available, one can compute and tabulate all the scattering rates in advance, store them, and just look them up as needed. For more details, see the book by Singh [16] and references therein, which treats MC method in transport in semiconductors.

For interacting particles, one can map to a similar problem as what was just mentioned. Instead of p processes, we now have Np processes. One of them must be picked at random. This is equivalent to picking a particle and a process at random. Let it scatter and update the state of the system (potentials and forces). In this case, the scattering rates may need to be computed as one goes on, and cannot be tabulated for additional use since they depend on the state of the system. This method suffers, however, from the approximation in which not all particles are moved at the same time. However, it can still be reasonable as the total rate, the denominator z, is now much larger and therefore the free flight time becomes much smaller, by a factor of N. This calculation becomes soon prohibitive as the number of particles is increased since the needed CPU is proportional to N. An alternative in this case, is to adopt the idea of MD, and fix the time step Δt such that it is at least an order of magnitude smaller than the smallest timescale (or largest transition rate), and move all particles at the same time. Since $w \, \Delta t$ is generally small, most of the time particles will have a "free flight" according to the external and mean field potentials. The way scattering is decided at each time step is to choose a random number in $[0,1)$, and make the following comparison:

$$\sum_{i=1}^{j-1} \Delta t \, w_i \leq r < \sum_{i=1}^{j} \Delta t \, w_i \, .$$

Note that the index i starts from unity: one only considers physical processes. Since $w \, \Delta t$ is chosen to be small, scattering events do not occur very often.

Conclusions and Discussion

The main concept behind various Monte Carlo simulation algorithms is based on random processes, as opposed to deterministic algorithms. Monte Carlo simulations generally follow the time dependence of a model for which a change does not proceed in rigorously predefined fashion, but rather in a stochastic manner. Success of such simulation algorithms depends on fast and efficient generation of a sequence of random numbers (or more often pseudo-random numbers) during the simulation.

Generally, in Monte Carlo simulations the most appropriate algorithm for a given simulation is specific to the case in hand. Algorithms already developed for simulation of macroscopic systems may not be valid for small systems of interest in nanotechnology.

Monte Carlo methods have found diverse applications in computational nanotechnology. The critical issue for appreciable advances in nanotechnology is the availability of fast and accurate simulation techniques in synthesizing and characterizing nanostructures. Considering the fact that a ten-nanometer cube may contain thousands of atoms and molecules there are significant challenges in using Monte Carlo simulation techniques to predict the characteristics and behavior of nano systems.

Bibliography

[1]. K. Esfarjani and G. A. Mansoori, *"Statistical Mechanical Modeling and its Application to Nanosystems"*, in *"Handbook of Computational Nanoscience and Nanotechnology"*, American Sci, Pub., Stevenson Ranch, CA, (to appear).

[2]. K. Ohno, K. Esfarjani and Y. Kawazoe, *"Computational Materials Science, From Ab initio to Monte Carlo"*, Springer series in Solid State, New York, NY, (1999).

[3]. D. E. Knuth, *"The art of computer programming"*, Vol. **2**, third edition, Addison-Wesley, Reading, MA, (1998).

[4]. D. J. Lehmer, *"Mathematical methods in large-scale computing units"*, Proc. 2nd Symp. on Large-Scale Digital Calculating Machinery, Cambridge, MA, (1949).

[5]. N. Metropolis, A. W. Rosenbluth, M. N. Rosenbluth, A. H. Teller and E. Teller, *J. Chem. Phys.* **21**, 1087, (1953).

[6]. W. H. Press, S. A. Teukolsky, W. T. Vetterling and B. R. Flannery; *"Numerical Recipes, The Art of Parallel Scientific Computing"*, Cambridge Univ. Press, Cambridge, MA, (1996).

[7]. D. M. Ceperley, *AIP Conf. Proc.* **690**, 1, 378, 25 Nov., (2003); D. M. Ceperley, *Rev. Mod. Phys.*, **67**, 279 (1995) and references therein.

[8]. K. Binder and D.W. Heerman, *"Monte Carlo Methods in Statistical Physics"*, 4th edition, Springer (2002); K. Binder, *"The Monte Carlo Method in Condensed Matter Physics"*, Vol. **71** in Topics in Applied Physics, Springer-Verlag (1992); K. Binder, *Applications of the Monte Carlo Method in Statistical Physics"*, Vol. **36** in Topics in Current Physics, Springer-Verlag, New York, NY, (1984).

[9]. M. E. J. Newman and G. T. Barkema, *"Monte Carlo Methods in Statistical Physics"*, Clarendon Press, Oxford, UK (1999).

[10]. D. P. Landau, K. K. Mon and H. B. Schuttler, *"Computer Simulation in Condensed Matter Physics IV"*, Springer, Berlin, Germany, (1992).

[11]. H. Gould and J. Tobochnik, *"An Introduction to Computer Simulation Methods"*, 2nd edition, Addison-Wesley, Boston, MA, (1996).

[12]. M. C. Mc. Millan, *Phys. Rev.* **138**, A442, (1964).

[13]. W. Krauth and O. Pluchery, *J. Phys. A*: Math Gen **27**, L715 (1994); W. Krauth, cond-mat/9612186 *"Introduction To Monte Carlo Algorithms"* in *Advances in Computer Simulation,* J. Kertesz and I. Kondor (Ed's), Lecture Notes in Physics

(Springer Verlag, 1998); W. Krauth, cond-mat/0311623 *"Cluster Monte Carlo algorithms"* Chapter of *New Optimization Algorithms in Physics*, A. K. Hartmann and H. Rieger (Ed's), (Wiley-VCh) (2004); J. M. Pomeroy, J. Jacobsen, C. C. Hill, B. H. Cooper and J. P. Sethna, *Phys. Rev. B* **66**, 235412 (2002).

[14]. R. H. Swendsen and J. S. Wang, *Phys. Rev. Lett.* **63**, 86, (1987).

[15]. M. A. Novotny, *Phys. Rev. Lett.* **74**, 1 (1995), Erratum: 75, 1424, (1995).

[16]. J. Singh, *"Electronic and Optoelectronic Properties of Semiconductor Structures"*, Cambridge Univ. Press, Cambridge, UK, (2003).

[17]. K. K. Bhattacharya and J. P. Sethna, *Phys. Rev. E* **57**, 2553, (1998).

[18]. A. Zangwill, "Physics at Surfaces", Cambridge Univ. Press, Cambridge, UK, (1988).

Chapter 5

Molecular Dynamics Simulation Methods for Nanosystems

"I do not know what I may appear to the world; but to myself I seem to have been only like a boy playing on the seashore, and diverting myself in now and then finding a smoother pebble or a prettier shell than ordinary, whilst the great ocean of truth lay all undiscovered before me." **Sir Isaac Newton**

Introduction

Alder and Wainwright originally introduced the computer-based molecular dynamics (MD) simulation method in the late 1950's [1,2]. Their study was on the behavior of hard-sphere fluid in macroscopic scale. Their important investigations produced a wealth of information about the role of intermolecular interaction energies on the macroscopic properties of fluids and the fact that the repulsive energies account for major part of the configurational properties of fluids. Their study also paved the way for the development of analytic hard-sphere equations of state and consequently the perturbation and variational theories of statistical mechanics for macroscopic systems [3-10].

Later in 1964 Rahman reported [11] results of his MD simulation of liquid argon using the Lennard-Jone intermolecular potential energy function which is considered an effective (realistic) potential energy function for simple fluids in macroscopic scale. While there were many other ongoing investigations on the applications of MD simulations since the original works of Alder and Wainwright, however, the first MD simulation of a realistic substance was reported by Rahman and Stillinger [12] in their simulation of liquid water in 1974.

In the interest of nanotechnology the first MD simulation of a nanostructure was studies into the dynamics and structure of protein by McCammon et al [13] which appeared in 1977. That involved the MD simulation of the bovine pancreatic trypsin inhibitor (BPTI). Today the method of MD simulation has gained popularity in many fields of science and engineering including quantum and statistical mechanics, material science, biochemistry and biophysics [14-19]. It serves as an important tool to study nanostructures such as protein structure determination and refinement; diamondoids, fullerene, and nanotube property studies; self-replication and self-assembly.

In the Molecular Dynamics (MD) simulation methods [14,15,20,21] the emphasis is on the motion of individual particles (atoms and/or molecules) within an assembly of N atoms, or molecules, that makes up the nano system under study. The dynamical theory employed to derive the equations of motion is either the Newtonian deterministic dynamics or the Langevin-type stochastic dynamics. The input data required are the initial position coordinates and velocities of the particles, in either a crystalline or an amorphous state, located in a primary computational cell of volume V.

To reduce the computer time, the simplifying assumption is made that each particle interacts only with its nearest neighbors, located in its own cell as well as in the image cells, within a specified cut-off radius. The $3N$ coupled differential equations of motion can then be solved by a variety of numerical finite-difference techniques as presented below.

Atoms and molecules in a solid or a cluster, except for hydrogen and helium, which have a relatively small mass, are considered as classical objects. Indeed, their thermal wavelength at temperature T, is

$$\Lambda = [h^2/2\pi m k_B T]^{1/2} = 1/\sqrt{M} \ A^\circ,$$

also known as deBroglie wavelength, where M is expressed in units of proton mass. For hydrogen, the deBroglie wavelength is $1.0\AA$, and for carbon it is less than $0.3\AA$ (at $0.0K$ temperature, one must consider the kinetic energy of the particle instead of Δt^2). One needs to compare this length to the size of the atom to decide whether or not to treat it classically. Therefore for small atoms, the zero point quantum motion is important, but for larger ones it is often neglected, and we use the classical Newtonian equations to describe their motion. Molecular Dynamics (MD) simulation is the method to numerically integrate the classical equations of motion, and to deduce atomic and molecular trajectories over some finite time scale in the order of nano or micro second at the most. From the trajectories, one can also obtain information about the dynamics of the atoms and molecules, can visualize a reaction, or compute mechanical and thermodynamic properties of a given system.

Principles of MD Simulation of Nanosystems

To simulate properties of matter in macroscopic scale through molecular dynamics, it is generally assumed that a limited number of the particles of matter are in a box with periodic boundary conditions. That means if the particles move out of the box from one side, we put them back in the box from the other side. This kind of computation will help to simulate macroscopic systems with the use of less computer time and to avoid surface effects due to the finite size of the box. The way this usually is done, in order to avoid using *IF statements* (which are more time consuming) in the FORTRAN CODE, is to use the following FORTRAN statement:

$$X(i)=X(i)-box*anint(X(i)/box),$$

where the $anint(x)$ function picks the nearest integer to x. By doing this, the homogeneous macroscopic system is replaced by a periodic box with particles moving exactly in the same manner in all the unit cells. Therefore, one important issue to consider is to handle fully the effect of long-range forces coming from particles in the neighboring boxes. To simulate properties of matter in nanoscale (like free cluster of molecules, a limited number of particles confined to a closed space, etc.) through molecular dynamics, there won't be any need for the assumption of periodic boundary conditions.

If particles are charged due to the ionic nature of the bonds, then long-range Coulomb interactions are present between the particles. The standard technique to handle this problem is to use the **Ewald sum method** [22] for summing the $1/R$ potential in three dimensions (see Glossary for more detail). But it can be generalized to $1/R^n$ in any dimension provided that the sums are convergent [23]. Other methods, which calculate the multipole moments of the potential for particles situated far away, are also used.

Other non-covalent interactions are usually treated by potential energy functions as presented in Chapter 2, which they all have algebraic decay. But in practice they are truncated by using a cut off distance and a shift as shown below:

$$W(r) = \phi(r) - \phi(r_{cut}) \qquad \text{for} \qquad r < r_{cut}, \qquad (1)$$

$$W(r) = 0 \qquad \text{for} \qquad r > r_{cut}, \qquad (2)$$

where $W(r)$ is the effective interaction potential energy function.

The number of operations for calculating the interparticle force on a given particle is, in principle, proportional to the total number of pair-interactions in the system if we ignore many-body interactions. Therefore, in general, we need to compute $N (N -1)/2$ interaction forces at every time step. In practice, due to the screening effect, each particle only sees the force due to its neighbors, and the task of computing all the forces becomes of the order of $c.N$, where c is the average number of nearest neighbors. Exceptions are long-range forces such as Coulomb interaction. In practice, many efforts, however complicated, are made to have an order of N algorithm in which case, it is always useful and necessary to have a neighbor list. Making a neighbor list is itself, unavoidably, large and has $N(N -1)/2$ terms, which is about $N^2/2$, or of the order of N^2, $O(N^2)$, when N is very large. This needs to be done only at the beginning of the simulation run. In every few time steps, it needs to be updated if one is simulating a liquid or a dense gas, but this task itself is of the order of N, $O(N)$, because one only needs to check which particles in the "second, third,...neighbor shells" entered the neighbor list of the considered particles. For this reason, to create a neighbor list, one needs to use a larger cutoff distance than that of the forces. Typically one can take as the cutoff distance of the neighbor list

$$r = r_{cutoff} + n.\Delta\theta.v_{max},$$

where n is the number of steps after which the neighbor list is updated and v_{max} is the largest typical speed of the particles.

For Coulombic or gravitational forces, which are long-ranged, the number of computational steps for forces is $N(N-1)/2$ which can be approximated to $N^2/2$ in large systems. For most nano system MD simulations there is no need for an approximation and one can deal with all the $N(N-1)/2$ computational steps. For large systems to make this task lighter, one can use what is called a particle-mesh method.

In MD simulation of nano systems each particle inside the system may be identified as a point. Then the space is discretized and the potential energy is calculated and represented on a grid. Since N is the number of grid points, the task of computation of the potential becomes of the order of $N^2/2$ or $O(N^2/2)$. For very large number of particles, when computational time is of concern the intermolecular potential energies could be calculated first in the Fourier space, in order to take advantage of the faster speed of Ewald sum method as described in Glossary.

Once the forces are known on all grid points, an interpolation can give their value at the actual particle positions. This treatment assumes that the potential due to Coulomb, or whatever long-range interaction we have, is a smooth function, which in effect is a field.

It is also possible to assign charges on each grid point, and avoid altogether the sum over particles $i=1,...,N$, and then solve a Poisson's equation. Charge assignment to a grid may be done by adding up the charges in the volume around that grid point. With free boundary conditions such as a gravitational problem, one can directly solve the Poisson's equation in real space. Discretized on a mesh of size M, it is equivalent to solving a linear system problem, which is $O(M^2)$. For macroscopic and other large systems with the need for periodic boundary conditions, one may use FFT to solve the Poisson's equation [1].

The particles, assumed to be like points, are moved according to the Newton's equation of motion:

$$m_i \frac{d^2 \vec{r}_i}{dt^2} = \vec{F}_i, \tag{3}$$

where the force F_i is coming from any external field plus the interparticle interactions. This simple form of the Newton's equation is applicable for an isolated system in which the total energy and total momentum are conserved. In statistical mechanical terms, Eq. (3) is applicable for a **microcanonical ensemble** simulation when the total energy of the multi-particle system are conserved. We will see later how to include the effect of a thermostat (heat bath) in the equations of motion for canonical simulations.

Integration of Newton Equation of Motion

To numerically integrate the equations of motion, one needs to discretize the time and, by using a Taylor expansion, write the position at a later time step as a function of the present position, velocity, acceleration and eventually higher derivatives of position as follows:

$$r(t + \Delta t) = r(t) + \Delta t \, v(t) + \frac{(\Delta t)^2}{2} a(t) + \frac{(\Delta t)^3}{6} b(t) + O(\Delta t^4),$$

$$v(t + \Delta t) = v(t) + \Delta t \, a(t) + \frac{(\Delta t)^2}{2} b(t) + O(\Delta t^3). \tag{4}$$

These are actually finite difference expressions for position and velocity.

Since the equation of motion is of second order, we need two initial conditions to start to integrate it for every particle. Usually the initial positions and velocities for particles are assumed to be at certain known values.

(i) With those initial values it is then possible to compute the forces acting on particles at time $t=0$.
(ii) Then by using the position equation above, truncated after the third term, we can update the position.

(iii) Afterwards, by truncating the velocity equation above after the second term we can update the velocity.

(iv) The acceleration, being proportional to the force, can then be computed exactly at every time step since positions are known.

From the above equations we see then that, with the truncations performed, the error in v is of the second order in Δt^2, and the error in r is of the third order in Δt.

Due to truncation of Eq.s (4) the b(t) terms and higher in position and velocity equations (third order term or higher in position and second order term or higher in velocity) are not computed and there is no use for them in the simulation. Of course, one could calculate them by writing a finite difference expression for the accelerations, but they are not necessary in simulation.

Of course, if Δt is not chosen carefully, the trajectories might diverge rapidly from their true paths. So we need appropriate algorithms to reduce the error in the time integration while keeping Δt as large as possible. The Verlet method [24], the Leap-Frog method [25], the Velocity-Verlet Method [26] and the predictor-corrector method [27] are four principal algorithms developed to help converge the MD simulations as they are presented below:

1. The Velet Method [25]: The simplest way to reduce the error in the time integration of the Newton's equation of motion while keeping Δt as large as possible is to expand the position at t-Δt as a power series in position, velocity and acceleration at time t and add the two equations (at times t-Δt and t+Δt) together to obtain the position at t+Δt :

$$r(t + \Delta t) = 2r(t) - r(t - \Delta t) + (\Delta t)^2 a(t) + O(\Delta t^4),$$
$$v(t) = [r(t + \Delta t) - r(t - \Delta t)]/2\Delta t + O(\Delta t^2).$$

$$(5)$$

This method, due to Verlet [24], works better, but can also become inaccurate due to the round-off errors resulting from subtraction of two large numbers and addition of a small number of the order of Δt^2. However, a preferred method is the Leap-Frog method due to Hockney and Eastwood [25].

2. The Leap-Frog Method [25]: In this method, the velocities are calculated at $n\Delta t + \Delta t / 2$, i.e. at half steps between the positions, whereas positions and accelerations are computed at multiples of Δt. This could also be a problem at the start of the run. But assuming we are given $v(t - \Delta t /2)$ and $r(t)$, we can first compute $a(t)$, then the velocity is computed at Δt time later, and next the position is calculated at time $t + \Delta t$ as follows:

$$v(t + \frac{\Delta t}{2}) = v(t - \frac{\Delta t}{2}) + \Delta t \, a(t),$$

$$r(t + \Delta t) = r(t) + \Delta t \, v(t + \frac{\Delta t}{2}). \tag{6}$$

Note that if we use periodic boundary conditions (as needed for MD simulation of macroscopic systems), and at $t + \Delta t$ the particle moves to the other side of the box. This will not be a problem in simulation of nano systems. It should be also pointed out that the Verlet algorithm will not give the correct velocity but the leap-frog does. Errors in the leap-frog method are of the order of Δt but the algorithm is more stable than that of the Verlet method.

3. The Velocity-Verlet Method [26]: Yet a more widely used method is called the Velocity-Verlet (VV) algorithm. Here, all positions, velocities and accelerations are computed at multiples of Δt. First the position is updated according to the usual equations of motion, then the acceleration is computed at that position, and then the velocity is computed from the average of the two accelerations. This algorithm is equally accurate for nano and macroscopic systems since the particles moving out of the box with periodic boundary conditions cause no error in calculations in this algorithm:

$$r(t + \Delta t) = r(t) + \Delta t \, v(t) + \frac{(\Delta t)^2}{2} a(t),$$

$$v(t + \Delta t) = v(t) + \frac{\Delta t}{2} [a(t + \Delta t) + a(t)]. \tag{7}$$

All the three variables (a, v and r) need to be defined and stored in this method. However, in terms of stability, this algorithm is superior to the Verlet and leap-frog algorithms because it is based on an **implicit** method; *i.e.* the velocity at $t + \Delta t$ is computed from the acceleration at t and at $t + \Delta t$. The superiority of this algorithm from the stability point of view is that it allows us to take a larger Δt to update positions, and with fewer steps, we can run longer simulations. As it is, after the update of the accelerations, the old accelerations are also needed. That makes *4 N* dimensional arrays to be stored instead of *3N* (*4* being the number of arrays of data multiplied by *N*, the number of particles). The way VV method is implemented in order to save memory is, first, to update the velocities $\Delta t / 2$ step later, *i.e.*,

$$v(t + \Delta t / 2) = v(t) + a(t)\Delta t / 2 , \qquad (8)$$

then positions are updated:

$$r(t + \Delta t) = r(t) + \Delta t \, v(t + \Delta t / 2) , \qquad (9)$$

and finally, from the new positions, new accelerations $a(t + \Delta t)$ are calculated and velocities are updated according to:

$$v(t + \Delta t) = v(t + \Delta t / 2) + a(t + \Delta t)\Delta t / 2 . \qquad (10)$$

If size of the memory is no objection, then this two-step update of velocities is unnecessary, and one can just use equation (7).

4. The Gear Predictor-Corrector Method [27]: To obtain more accurate trajectories with the same time step, or to be able to use larger time steps, one can use a higher order method called Gear Predictor-Corrector (GPC). Assuming that the trajectories are smooth functions of time, in this method, the information on the positions of 2 or 3 previous time steps is also used. In principle, at first the positions, then velocities and finally accelerations are predicted at $t + \Delta t$ from the finite difference formula up to the order n. Below the related third order formulas are reported,

$$r^P(t + \Delta t) = r(t) + \Delta t\, v(t) + \frac{(\Delta t)^2}{2} a(t) + \frac{(\Delta t)^3}{6} b(t) + O(\Delta t^4),$$

$$v^P(t + \Delta t) = v(t) + \Delta t\, a(t) + \frac{(\Delta t)^2}{2} b(t) + O(\Delta t^3), \tag{11}$$

$$a^P(t + \Delta t) = a(t) + \Delta t\, b(t) + O(\Delta t^2).$$

But after the position is predicted at $t + \Delta t$, we can compute the actual correct acceleration, a^C, at that time. The error we have caused in predicting the acceleration from the above formula is: $\Delta a = a^C - a^P$. Using the perturbation theory, one can argue that the error on the predicted positions and velocities is also proportional to Δa, and hence, we can correct for that error by using the following equations:

$$r^C(t + \Delta t) = r^P(t + \Delta t) + c_0 \Delta a,$$

$$v^C(t + \Delta t) = v^P(t + \Delta t) + c_1 \Delta a, \tag{12}$$

$$a^C(t + \Delta t) = a^P(t + \Delta t) + c_2 \Delta a,$$

where $c_2 = 1$, and the other coefficients may be determined by the method reported in [27] in order to achieve the highest possible accuracy. In a third-order predictor-corrector method, $c_0 = 0$, $c_1 = 1$, $c_2 = 1$, and in a fourth-order predictor-corrector method, $c_0 = 1/6$, $c_1 = 5/6$, $c_2 = 1$, $c_3 = 1/3$. Higher-order predictor-corrector methods can also be used, but they require more computer memory and can consume more CPU without appreciably improving the accuracy [21].

Choice of the Time Increment Δt [20]: The important parameter to choose in an MD simulation is the time increment Δt. In a microcanonical ensemble simulation, the total energy of the system must be conserved. If Δt is too large, steps might become too large and the particle may enter the classically forbidden region where the potential energy is an increasing function of position. This can occur when two particles collide or when a particle hits the "wall" imposed by the external potential. Entering the classically forbidden region means that the new potential energy has become higher than the maximal value it is allowed to have. In this case, the total energy has increased, and this phenomenon keeps occurring for large step sizes until the total energy diverges. So, depending on the available total energy, Δt should be chosen small enough so that the total energy remains constant at all times, but not too small that it would require an extremely large number of steps to perform the simulation. The optimal value of Δt is usually found by trial and error. One femto (10^{-15}) second is a good trial guess, but the optimal value really depends on the initial energy and the kind of potential considered.

MD Simulation of Systems in Contact with a Heat Bath: Thermostats

For a large class of problems in the physics and chemistry of nano systems, the type of system that is considered is the closed (control mass) system. This is a system with a fixed volume, V, a fixed number of particles, N, maintained at a constant temperature, T. Within statistical mechanics, such a system is represented by a constant (NVT), or canonical ensemble [28], where the temperature acts as a control parameter.

In canonical ensemble simulations, the total energy is not a conserved quantity, rather, the temperature T is a constant, and particles exchange energy with a heat bath external to the system, so that, for example the kinetic energy per particle of monatomic particles remains, on average, equal to $(3/2)k_BT$. Modeling this energy exchange mechanism, or a constant-temperature MD simulation, can be done in many different ways, three of which are discussed below:

1. Velocity Scaling Thermostat [29]: Since the average kinetic energy is $(3/2)k_BT$ per monatomic particle, one can simply force the total kinetic energy to be $(3/2)Nk_BT$ for N monatomic particles which comprise the system at all times. Therefore we can multiply all velocities by a scale factor to make the kinetic energy equal to $(3/2)Nk_BT$. This is the simplest way that the contact with a heat bath at temperature T can be modeled. But one may argue that this is not physically correct since the kinetic energy is a fluctuating quantity and only its average value for N monatomic particles is equal to $(3/2)Nk_BT$.

If one is simulating rigid multi-atomic molecules, which possess more than 3 degrees of freedom per particle, the equipartition theorem must be applied carefully. One has then $(3/2)k_BT$ for the translation of the center of mass, and $(3/2)k_BT$ for the rotation around the center of mass per each molecule (k_BT only for diatomic molecules). The kinetic energy of $3k_BT$, or $(5/2)k_BT$ for diatomic molecules, must be shared by all the atoms in the molecule. Even if the assumption of rigidity is absent, velocities must be scaled such that the center of mass has kinetic energy of $(3/2)k_BT+(3/2)k_BT$ for rotations and translations, and $(3/2)(N-2)k_BT$ for the vibrational (internal) degrees of freedom. It means that velocities of the particles need to be decomposed to center of mass + relative velocity. Then each component is separately scaled, and then added back to obtain the absolute speed of the particles.

2. The Nose-Hoover Extended-System Thermostat [30-33]: A method that generates the canonical ensemble distribution in both the configuration and momentum parts of the phase space was proposed by Nosé [30-32] and Hoover [33] and is referred to as the extended-system method. According to this method, the simulated system and a heat bath couple together to form a composite system. This coupling breaks the energy conservation that otherwise restricts the behavior of the simulated system and leads to the generation of a canonical ensemble. The conservation of energy still holds in the composite system, but the total energy of the simulated system is allowed to fluctuate.

For the Nose-Hoover thermostat, there is a single degree of freedom, ξ, which is interacting with the particles. The particle Lagrangian is modified to include also the thermostat. One can then deduce the coupled equations of motions for the particles and thermostat. In this case the

Helmholtz free energy of the whole system of particles plus thermostat is a conserved quantity. The equations of motion in this scheme are:

$$\frac{d^2\vec{r}_i}{dt^2} = \frac{\vec{F}_i}{m_i} - \xi\frac{d\vec{r}_i}{dt},$$

$$\frac{d\xi}{dt} = \frac{1}{\tau^2}[\sum_{i=1,N}(m_iv_i^2/f\,k_BT)-1],$$

(13)

where f is the total number of degrees of freedom, and τ is a free parameter modeling the strength of the coupling of the system to the thermostat. A small value of τ yields a large derivative for ξ and hence a large additional friction force in the motion of the particles. Small τ then means strong coupling to thermostat. The parameter τ has the dimensions of "time" and thus can be called a "relaxation time". The parameter τ can be taken to be about one pico (10^{-12}) second but, its exact value should be adjusted depending on how and what damping process one desires to model. The Hamiltonian from which these equations of motion can be derived is:

$$H = \sum_i\frac{1}{2}m_iv_i^2 + \sum_{i<j}v(r_{ij}) + (\frac{1}{2}\tau^2\xi^2+\xi)f\,k_BT.$$

In the numerical integration of the modified Newton equation of motion, Eq. (13), one can adjust the time increment Δt so that the above Hamiltonian is a conserved quantity. Note that if the total kinetic energy per monatomic particle is just $(3/2)k_BT$, then ξ is a constant that we can set initially to be zero. If the term,

$$(\frac{1}{2}\tau^2\xi^2+\xi),$$

in Hamiltonian is positive, then ξ has to be positive indicating of the existence of a friction force which slows down the particles. And if this term is negative, ξ has to be a negative value, and pushes the particles further along their momentum so as to increase their kinetic energy. One can see that this coupling tries to keep the kinetic energy of the particles around $(3/2)k_BT$. This thermostat can be used for an *NVT* canonical simulation.

If one wishes to perform an *NPT* simulation with a variable box volume $V(t)$ but constant pressure P_o, it is also possible to add an additional "barostat" degree of freedom coupled to the system. The resulting equations are then:

$$\frac{dr}{dt} = v(t) + \eta(t)[r(t) - R_o],$$

$$\frac{dv}{dt} = \frac{F}{m} - [\eta(t) + \chi(t)]v(t),$$

$$\frac{d\chi}{dt} = \frac{1}{\tau_T^2}[\sum_{i=1,N} (m_i v_i^2 / f k_B T - 1],$$ (14)

$$\frac{d\eta}{dt} = \frac{(P - P_o)V(t)}{\tau_P^2 N k_B T},$$

$$\frac{dV}{dt} = 3\eta(t)V(t),$$

where τ_P is the pressure relaxation time, P_o is the target pressure, R_o is the center of mass coordinate, and V is the system volume, also considered here as a dynamical variable. In this case, the Gibbs free energy G is the conserved quantity. It is given by.

$$G_{NPT} = U_{NVT} + P_0 V(t) + f k_B T \tau_p^2 \eta^2(t)/2,$$

as proposed by Melchionna and co-workers [34]. Conserved quantities in general have the advantage of the possibility of checking the stability of the update algorithm, and determining the relevant parameters such as Δt.

3. The Langevin Thermostat [30-33]: Yet a third method to exchange energy between the heat bath and the particles is to use the Langevin thermostat. In this case the particle trajectories are modified by two additional forces: a friction force and a random force of zero time average, with white noise of mean squared deviation equal to $2m\,k_B T\,/\tau\,\Delta t$ per degree of freedom for monatomic particles:

$$m\frac{d^2 x}{dt^2} = F_x - \frac{m}{\tau}\frac{dx}{dt} + f;$$

$$<f> = 0; \tag{15}$$

$$<f(t)f(t')> = \delta(t-t')2mk_B T/\tau.$$

Note that when implementing this algorithm, the Dirac delta function must be replaced in the variance calculation by $1/\Delta t$. So, in practice, one generates random numbers with Gaussian distribution of

$$Mean=0 \text{ and } variance = 2m\,k_B T\,/\tau\,\Delta t\ ,$$

for each direction of the force. One simple way of generating random numbers of Gaussian distribution is the following [21]:

- Generate two random numbers a and b in $[0,1)$.
- Calculate

$$A = M + \sigma\sqrt{-2\ln a}\,\cos 2\pi b \quad \text{and} \quad B = M + \sigma\sqrt{-2\ln a}\,\sin 2\pi b\ .$$

The numbers A and B have, both, a common Gaussian distribution of mean M and variance σ. In this thermostat, τ is the relaxation time and represents the mean collision time between the real particles of the system and some fictitious (virtual) particles, which collide at random with them, and are represented by the two additional forces. This thermostat could be interesting since it physically models some stochastic collisions with virtual particles. It can be shown, that monatomic particles following this dynamics, will have an average kinetic energy of $(3/2)k_B T$ per particle in three dimensions. The collision rate $1/\tau$ models the strength of the interaction with the thermostat. Large τ means weak friction and weak random force, i.e. weak interaction with the bath since particles do not collide so often with the virtual bath particles. A reasonable value for τ would be about 1000 time increment Δt, but it must be adjusted based on the kind of thermal contact one wants to model. Again, some trial and error is necessary to adjust this parameter.

In Figures 1 and 2, we can see the time evolution of the kinetic energy to $(3/2)k_BT$ as a function of time for several values of the relaxation time. The initial configuration is a deformed *13*-atom cluster of icosahedral shape. Three values of τ were chosen for this illustration.

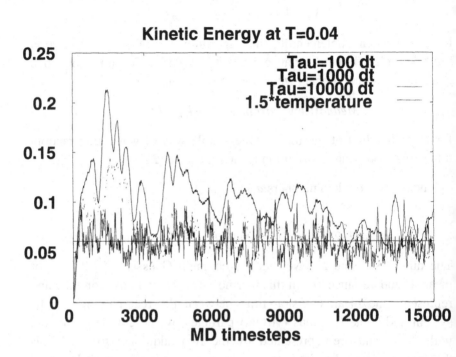

Figure 1. Time evolution of the kinetic energy of a 13 atom-argon cluster for 3 different relaxation times $\tau = 100\,\Delta t; 1000\,\Delta t; 10000\,\Delta t$. It can clearly be seen that the blue curve corresponding to 10000dt relaxes to the value 0.06 after about 10000 or more steps, and the green curve needs about 3000 steps to relax [20].

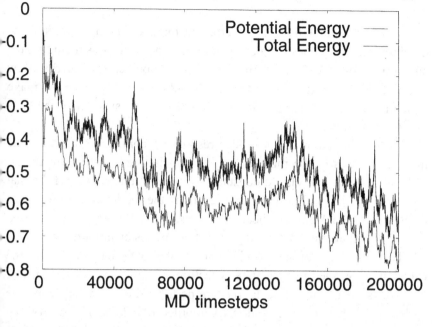

Figure 2. Evolution of the total and potential energy of a 13 atom argon cluster at T=0.08. After about 200,000 steps [20].

The Langevin thermostat with a small relaxation time would also be a useful tool for performing structural relaxations. Indeed the particles move along the steepest descent direction (i.e. along the direction of forces F) but are also given some random momentum to escape eventual potential barriers. A small value of τ would be better in relaxation calculations. It can eventually be increased towards the end of the run. It is also possible to lower the temperature towards the end of the run just as it is done in simulated annealing as presented later in this chapter.

Calculations Resulting from MD Simulations

From the MD simulation of trajectory of the particles, it is possible to obtain numerical values for a number of response functions including the autocorrelation function of the positions, velocities, or energy flow (particle energy times its velocity), in addition to be able to make animation of the dynamics of the particles in the system to see how things move in nanoscale.

From application of the linear response theory, one can calculate many properties of the molecular system including the structure factor, diffusion coefficient, elastic constants, phonon frequencies, and thermal conductivity in addition to equilibrium thermodynamic properties (pressure, internal energy, heat capacities, etc.). We can compute and predict all the above mentioned physical properties quantitatively.

The other important application of the MD method is in the calculation of the ground state structure of molecules and clusters. Nowadays, many such calculations are performed on very large organic molecules such as proteins and DNA in order to understand their folding and unfolding mechanisms and some of the reactions, which take place on small time scales. In fact some of these reactions may occur in milliseconds up to seconds. With today's computers it is impossible to run such long time-scales, considering that a typical MD time step is of the order of a femto (10^{-15}) second. The structure of the grains in granular materials, or clusters adsorbed on surfaces or simply in the gas phase, is also of importance for obvious technological reasons.

Conclusions and Discussion

From the above presentations it is obvious that the MD simulation methods are now strong means of simulating nano-scale models of matter. Since mid-1970s MD simulation has found applications in the study of static and dynamic properties of nanostructures. MD simulation is also considered as a technique for obtaining structural information that is complementary to property calculations. In recent years MD simulation methods have attracted a great deal of attention from nanotechnology researchers with diverse interest and backgrounds.

An MD simulation consist of the numerical solution of the Newton's equations of motion of a system consisting of an atom, a molecule or an assembly of atoms and molecules to obtain information about its time-dependent properties. Because of its nature MD simulation is an ideal technique for prediction of the behavior of nano systems consisting of finite number of particles. It is through MD simulation that one can relate collective dynamics of a finite number of particles accurately to single-particle dynamics. For example, how molecules are formed as a result of covalent bonds between atoms, how do molecular building blocks could self-assemble due to non-covalent interactions to produce materials with preferred properties, and many other important questions. Today in the nanotechnology related literature, one can find MD simulation of chain molecules self-assembly into structures such as micelles, vesicles, and lamellae, MD simulation of solvated proteins, protein-DNA complexes, antibody-diamondoid coupling studies, addressing a variety of issues of interest. The large number and variety of the existing free-domain and commercial simulation packages is an indication of its important role in science and technology. Currently numerous specialized MD simulation techniques for various particular problems exist. They include, for example, mixed quantum mechanical - classical MD simulations, which are being employed to study enzymatic reactions in the context of the full protein. MD simulation techniques are also widely used in experimental procedures such as NMR structure determination and X-ray crystallography.

Bibliography

[1]. B. J. Alder and T. E. Wainwright, *J. Chem. Phys.* **27**, 1208, (1957).

[2]. B. J. Alder and T. E. Wainwright, *J. Chem. Phys.* **31**, 459, (1959).

[3]. M. S. Wertheim, *Phys. Rev. Letts*, **10**, 321, (1963).

[4]. E. Thiele, *J. Chem. Phys.* **30**, 474, (1963).

[5]. N. F. Carnahan and K. E. Starling, *J. Chem. Phys.* **51**, 635, (1969).

[6]. J. L. Lebowitz, *Phys. Rev.* **133**, A895, (1964).

[7]. G. A. Mansoori, N. F. Carnahan, K. E. Starling and T. W. Leland, *J. Chem. Phys.*, **54**, 4, 1523, (1971).

[8]. J. A. Barker and D. Henderson, *Rev. Modern Phys.*, **48**, 587, (1976).

[9]. E. Z. Hamad and G. A. Mansoori, *Fluid Phase Equil.*, **37**, 255, (1987).

[10]. M. Mohsen-Nia, H. Moddaress and G. A. Mansoori, *Chem. Eng. Comm.* **131**, 15, (1995).

[11]. A. Rahman, *Phys. Rev. A*, **136**, 405, (1964).

[12]. F. H. Stillinger and A. J. Rahman, *Chem. Phys.* **60**, 1545, (1974).

[13]. A. McCammon, B. R. Gelin, and M. Karplus, *Nature*, **267**, 585, (1977).

[14]. J. M. Haile, *"Molecular Dynamics Simulation - Elementary Methods"* John Wiley & Sons, Inc., New York, NY, (1992).

[15]. D. C. Rapaport, *"The Art of Molecular Dynamics Simulation"*, Cambridge Univ. Press, Cambridge, UK, (2004).

[16]. J. A. McCammon and S. C. Harvey, *"Dynamics of Proteins and Nucleic Acids"*, Cambridge Univ. Press, Cambridge, UK, (1987).

[17]. D. Frenkel and B. Smit *"Understanding Molecular Simulation.* Academic Press, New York, NY (2001).

[18]. O. M. Becker, A. D. Mackerell Jr, B. Roux and M. Watanabe, *"Computational Biochemistry and Biophysics"*, Marcell-Dekker, Inc., New York, NY, (2001).

[19]. T. Schlick, *"Molecular Modeling and Simulation"* Springer, New York, NY, (2002).

[20]. K. Esfarjani and G. A. Mansoori, *"Statistical Mechanical Modeling and its Application to Nanosystems"*, *Handbook of Theoretical and Computational Nanoscience and Nanotechnology*, American Sci. Pub., Stevenson Ranch, CA (to appear), (2004).

166

[21]. M. P. Allen and D. J. Tildesley, *"Computer Simulation in Chemical Physics"*, Nato ASI Series C, Vol. 397, Kluwer Acad. Pub., New York, NY, (1992).

[22]. P. Ewald, *Ann. Phys.* **64**, 253, (1921).

[23]. K. Ohno, K. Esfarjani and Y. Kawazoe, *"Computational Materials Science, From Ab initio to Monte Carlo"*, Springer series in Solid State, New York, NY, (1999).

[24]. L. Verlet "Computer Experiments on Classical Fluids. I. Thermodynamical Properties of Lenard-Jones Molecules." *Phys. Rev.* 159, 98-103, (1967).

[25]. R. W. Hockney and J. W. Eastwood; *"Computer Simulation Using Particles"*, McGraw-Hill, (1981).

[26]. W. P. Swope, H. C. Anderson, P. H. Berens and K. R. Wilson, J. Chem. Phys., **76**, 637, (1982).

[27]. C. W. Gear; *"Numerical Initial Value Problems in Ordinary Differential Equations"*, Prentice Hall, Englewood Cliffs, NJ, (1971).

[28]. R.K. Pathria, *"Statistical Mechanics"*, Pergamon Press, Oxford, UK, (1972).

[29]. H. J. C. Berendsen, J. P. M. Postma, W. F. van Gunsteren, A. DiNola, and J. R. Haak, *J. Chem. Phys.*, **81**, 8, 3684, (1984).

[30]. S. Nosé, *Mol. Phys* **52**, 255, (1984).

[31]. S. Nosé, *J. Chem. Phys.* 81, 511, (1984).

[32]. S. Nosé, *Prog. Theor. Phys. Suppl.* **103**, 1, (1991).

[33]. W. G. Hoover, *Phys. Rev. A* **31**, 1695, (1985).

[34]. S. Melchionna, G. Ciccotti, and B. L. Holian, *Molec. Phys.* **78**, 533, (1993).

Chapter 6

Computer-Based Simulations and Optimizations for Nanosystems

"... augmented computational capability now enables sophisticated computer simulations of nanostructures. These new techniques have sparked excitement in nearly all parts of the scientific community. Traditional models and theories for most material properties and device operations involve assumptions leading to "critical scale lengths" that are frequently larger than 100 nm. When the dimensions of a material structure is under the respective critical length scale, then the models and theories are not able to describe the novel phenomena. Scientists in all materials and technology disciplines are in avid pursuit of the fabrication and measurement of nanostructures to see where and what kind of interesting new phenomena occur. Further, nanostructures offer a new paradigm for materials manufacture by assembling (ideally utilizing self-organization and self-assembly) to create an entity rather than the laborious chiseling away from a larger structure". **Richard E. Smalley**

Introduction

As we enter into the new century it is probably as good a time as any to look ahead and try to glimpse future trends in our society. With the abundance of powerful personal computers and workstations as well as plentiful supercomputer "time" available to researchers, the role of computational methods in the advancement of science and technology has become quite important. Such enhanced computational facilities have also brought about the advancement of nanoscience and nanotechnology [1]. These advancements will most likely continue and the tendency will be towards increased utilization of numerical models and high-tech imaging techniques. There will be great incentives to predict nanotechnology-related possibilities, structures, processes and events in order to understand nano systems in terms of their achievabilities and interfaces. This will allow us to enter the nanotechnology era faster than expected due to the advent of computational nanotechnology through which nanostructures and nanoprocesses could be accurately modeled and their behavior could be predicted [1].

Computer-based molecular simulations and modeling are the one of the foundations of computational nanotechnology. Due to the high cost of laboratory facilities necessary and slowness of the experimentation process it may not be wise to rely on experimental nanotechnology alone. Some refer to modeling and simulations as computational experimentations. This is because the computational investigator has control over the forces that could influence the structure, properties and dynamics of nano systems and the ability to perform a systematic study of the actions and reactions of each forc. The accuracy of such computational experimentations will depend on the accuracy of the intermolecular interactions as well as the numerical models and simulation schemes used. Provided the accuracy of the computational scheme is guaranteed one can use that to investigate various nonlinear interactions, the results of which could be completely unexpected and unforeseen. Numerical modeling and computer-based simulations will also help to advance the theoretical part of the nanoscience, which in turn will allow us to develop useful analytic and predictive models as it has been the case for macroscopic systems.

The most widely used methods in computational nanotechnology are computer-based simulation techniques. The two frequently used simulation approaches are Monte Carlo (MC) and Molecular Dynamics (MD) methods, which were presented, in the previous two chapters. In this chapter we categorize all the various simulation methods resulting from the basic MC and MD methods. We will also introduce the optimization techniques needed in MC and MD simulations.

A. Classification of Simulation Methods Based on Accuracy and Computational Time

The computer-based simulation methods, being developed for nano systems, generally consist of a computational procedure performed on a limited number of atoms, molecules, molecular building blocks or macromolecules confined to a limited, but small, geometrical space [1].

Generally, the cell in which the simulation is performed could be replicated in all spatial dimensions, generating its own periodic images containing the periodic images of the original N atoms or molecules. This is the periodic boundary condition, and is introduced to remove the undesirable effects of the artificial surfaces associated with the finite size of the simulated system. The forces experienced by the atoms and molecules are obtained from a prescribed two-body, or many-body, interatomic or intermolecular potential energy function, $\Phi_I(r_{ij})$, as discussed in Chapter 2 and according to

$$F_i = - \sum_{j>i} \nabla_{ri} \Phi_I(r_{ij}) , \qquad (1)$$

where r_{ij} is the separation distance between the two particles i and j.

The most important input to such computation is the interparticle energy / force function for interaction between the entities composing the nano system. Accurately administered computer simulations can help in three different ways:

(i) They can be used to compare and evaluate various molecular-based theoretical models. For this purpose we can use the same intermolecular potential function in the simulation as well as in the theoretical calculation.

(ii) They can help evaluate and direct an experimental procedure for nano systems.

(iii) An ultimate use of computer simulation is its possible replacement of an experiment which otherwise may not be possible with the present state of the technology or may be too costly, but provided accurate intermolecular potentials are available to be used in the development.

Methods used for simulation of various properties of nano scale systems differ in their level of accuracy and the computer time necessary to perform such calculations. Accordingly, the required computer-time-scale for these methods can be from tens of picoseconds for an *ab initio* molecular dynamics calculation to few microseconds or more for classical molecular dynamics simulation. There are also computer simulations requiring very long computer times, such as aggregation or cluster growth. In this case we may need to use supercomputers to achieve fast results.

There is extensive number of references on the subject; however in what follows, we will mainly give reference to sources which include a more general treatment of the subject. More specific references can be found in these sources.

Methods with the highest degree of accuracy (very CPU-intensive)

- Input: atomic species and their coordinates and symmetry of the structure; eventually interaction parameters (for model calculations).
- Output: Total energy, excitation energies, charge and spin densities, forces on atoms.
- Purpose: Investigation of both electronic and atomic ground state, optical and magnetic properties of weakly interacting and also strongly interacting correlated systems.

Examples:

Ab initio methods for electronic structure calculation of correlated systems [2-4].

Quantum Monte Carlo: Variational, fixed-node, Green's function and path-integral [3,5].

Quantum Chemistry: Configuration Interaction [6,7].

Many-body: GW [8], Coupled-Cluster [9].

Methods with the second highest degree of accuracy

- Input: atomic species and their coordinates and symmetry of the structure; eventually pseudopotentials or Hamiltonian parameters for the species considered.
- Output: Total energy, Charge and spin densities, forces on atoms, electron energy eigenvalues, capability of doing Molecular Dynamics (Car-Parinello MD of short timescale phenomena < 100 *ps*), vibrational modes and phonon spectrum.
- Purpose: Accurate calculation of ground state structure by local optimization; calculation of mechanical, magnetic and optical properties of small clusters and perfect crystals of weakly interacting electron systems; estimation of reaction barriers and paths.

Examples:

> *Ab initio* methods for normal Fermi liquid systems (one electron theories) based on either Hartree-Fock (chemistry) [4,6,7,10] or Density functional (Physics) theories [11-15].

Semi-empirical methods

- Input: atomic species and their coordinates; parameters of the interparticle potential, temperature and parameters of the thermostat or other thermodynamic variables.
- Output of Tight-Binding (TB) (still quantum mechanical): Total energy, Charge and spin densities, forces on atoms, particle trajectories (capability of doing Molecular Dynamics: timescales of up to nanoseconds or more), phonon calculation; mechanical, magnetic and optical properties (approximate) of clusters and crystals.
- Output of classical potentials, faster than TB and easily parallelizable: Total energy, forces on atoms, particle trajectories (capability of doing Molecular Dynamics: time scales of up to a micro second), phonon calculation; mechanical properties (approximate) of large clusters and imperfect crystals (still using periodic boundary conditions).

- Purpose: Search for ground state structure by Genetic Algorithms (GA), Simulated Annealing (SA) or local optimization if a good guess for the structure is known; simulation of growth or some reaction mechanisms; calculation of response functions (mostly mechanical or thermal properties) from correlation functions; thermodynamic properties.

Examples:

Semi-empirical methods for large systems or long time scales [16,17].

TB or LCAO (quantum mechanical) [17, 18].

Molecular dynamics based on classical potentials or force fields [16, 19, 20].

Stochastic methods

- Input: parameters of the interparticle potential, temperature and parameters of the thermostat or other thermodynamic variables.
- Output: statistics of several quantities such as energy, magnetization, atomic displacements ...
- Purpose: Investigation of long timescale non-equilibrium phenomena such as transport, growth, diffusion, annealing, reaction mechanisms and also calculation of equilibrium quantities and thermodynamic properties (all with approximate and simple interparticle potentials).

Examples:

Monte Carlo (walk towards equilibrium) [5, 21-23].

Kinetic or dynamical Monte Carlo (growth and other non-equilibrium phenomena) [24-26].

In what follows, we will discuss the two non-quantum methods used to deal with the dynamics of atoms in nano scale systems (although interatomic forces may be computed quantum mechanically):

The first part of this report introduces the popular Monte Carlo (MC) method, which uses random numbers to perform calculations. MC is used in many different areas, which include, but not limited to,

(i) estimation of large-dimensional integrals,
(ii) generating thermodynamic ensembles in order to compute thermal averages of physical equilibrium quantities of interest,
(iii) simulation of non-equilibrium phenomena such as growth and computation of distribution functions out of equilibrium, known as Kinetic Monte Carlo.

The second part concerns the Molecular Dynamics (MD) method, which simply deals with predicting the trajectories of atoms subject to their mutual interactions and eventually an external potential. MD is used in many different areas, which include, but not limited to,

(i) computation of transport properties such as response functions: viscosity, elastic moduli, thermal conductivity,
(ii) thermodynamic properties such as total energy and heat capacity,
(iii) dynamical properties such as phonon spectra.

MD deals with atomic phenomena at any time scale from femto to pico, nano or even micro seconds. This section is ended with the introduction of optimization (local and global) techniques.

A comparison between the two methods of MC and MD in ground state optimization of clusters can be found in [26].

B. Classification of Optimizations in Molecular Simulations

Generally, optimization in any numerical computation is to seek an answer for the question "What is the best value of a multi-variable function among the many numbers generated during the computation? [27] Such questions arise routinely in molecular simulations where a complex system is described by an assembly of N particles in a $6N$-dimensional phase space. In such computations a cost or energy function is associated with every configuration. The objective is to seek sets of data points that maximize (or minimize) the objective function.

There are two kinds of optimizations required in every molecular simulation, which are called local optimization and global optimization which are classified according to the features of the target problem. [28,29].

In each of these categories there are many optimization techniques and several commercial packages available for use.

Local Optimization Methods [27-29]

The simplest of the optimization methods, which may date back two centuries, or even more, since they carry Newton's name, deal with local optimization. A particle is inside a not necessarily harmonic well in an N-dimensional space. N is the number of degrees of freedom typically equal to three times the number of particles. We would like to take the particle to the minimum of that well as fast as possible by using some dynamics, not necessarily Newtonian. The landscape is defined by the "cost function" $E(x)$ which, in our case of interest, is the potential energy of the system.

1. Steepest Descent Method (SDM): The simplest optimization method that comes to mind is just to move along the force applied on the particle. This is called the steepest descent method (SDM) since the force is minus the gradient of the potential energy, which we want to minimize. This simple algorithm is as follows:

$$x(t + \Delta t) = x(t) + \lambda f[x(t)], \qquad (2)$$

where λ is a positive constant coefficient, and $f(x) = -\nabla E(x)$ is the force acting on the particle at x. Eq. (2) is a discretized version of

$$dx/dt = \lambda f(x).$$

Here one has to think of x and f as N-dimensional arrays, and the time variable t is discretized and represents the minimization variable. It is better to write Eq. (17) as mathematicians do:

$$x(i + 1) = x(i) + \lambda f[x(i)],$$

where i is the iteration number. This method certainly works because at each step the potential energy is lowered, but its convergence is often quite slow. It could become extremely slow if the valley around the minimum is quite anisotropic. The particle then will keep oscillating around the valley, which leads down to the minimum, and it would take many steps before reaching the minimum.

2. Damped Newtonian Dynamics Method: Instead of the standard SDM one may also adopt damped Newtonian dynamics,

$$d^2x/dt^2 = \lambda f(x) - 2\mu \, dx/dt, \qquad (3)$$

which also falls in the category of the SDM, but can be faster than SDM. In its discretized form, it can be written as:

$$x(i + 1) = [2x(i) + \lambda f(x(i)) - x(i - 1)(1 - \mu)]/(1 + \mu), \qquad (4)$$

where (λ, μ) are two appropriately chosen independent parameters.

3. Conjugate Gradients Method (CGM): A more intelligent method is called the conjugate gradients method (CGM). This method will find the minimum of the anisotropic valley around the minimum much faster than the SDM. In principle, it finds the minimum of a quadratic function in N dimensions in exactly N line minimization steps. Provided that one is near the minimum, the convergence to the minimum is almost as fast the SDM since the function is "almost quadratic".

For the sake of completeness and brevity we will just mention the algorithm to implement and will bypass its explanation. Interested readers can consult specialized books on the subject [30].

Step (1). Starting from a point $x(0)$; perform a line minimization along the steepest descent direction; find λ such that

$$x(1) = x(0) + \lambda f(0); \tag{5}$$

minimizes the energy assuming

$$c(0) = f(0). \tag{6}$$

Step (2). Calculate the next SDM direction at that point,

$$f(i+1) = f[x(i+1)] = -\nabla E[x(i+1)] \tag{7}$$

Step (3). Calculate the conjugate direction

$$c(i+1) = f(i+1) + b_i\, c(i) \tag{8}$$

where

$$b_i = \frac{\|f(i+1)\|^2}{\|f(i)\|^2}. \tag{9}$$

Step (4). Update x by doing a line minimization,

$$x(i+1) = x(i) + \lambda_i\, c(i). \tag{10}$$

This defines λ_i

Step (5). GO TO Step (2) until the calculation converges, meaning the change in the cost function E becomes less than a tolerance value.

4. Quasi-Newton Methods: Newton's method is an algorithm to find the zero of a function f in one dimension. The expression for the iterative scheme is as follows:

$$x(i+1) = x(i) - f(x(i)) / f'(x(i)) \tag{11}$$

In our minimization problem, we are also looking for the zero of the force vector. That is where the energy minimum is. This method can easily be generalized to N dimensions. The only difference with the other methods of minimization being that the derivative of the vector f, the Hessian, becomes an NxN matrix, and furthermore, it needs to be inverted and then multiplied by the force vector. This computation is very difficult and demanding. The computation of the matrix elements of the Hessian is quite computer-time-consuming and complicated; its inversion is also a very CPU intensive task. Therefore, one resorts to quasi-Newton methods, which try to approximate the Hessian as the iterations proceed. These methods only need the forces (or gradients of the potential energy) as input. They start with the steepest descent method, which approximates the Hessian by the identity matrix times a constant. Then at each step, using the information on all previous forces and positions, improve their guess of the inverse Hessian matrix. They are also known as Broyden's method [31] since Broyden was the first to propose this method. The basics of the Quasi-Newton algorithm is as the following [31,32]:

Step *1:*
$$x(2) = x(1) - \lambda f(1). \tag{12}$$

Step *i:*
$$x(i+1) = x(i) - \lambda f(i) - \sum_{j=2}^{i} u(j) < v(j) \,|\, f(i) >. \tag{13}$$

where,

$$u(j) = -\lambda(f(j) - f(j-1)) + x(j) - x(j-1)$$

$$- \sum_{l=2}^{j-1} u(l) < v(l) \,|\, f(j) - f(j-1) >,$$

and

$$v(j) = \frac{f(j) - f(j-1)}{< f(j) - f(j-1) \,|\, f(j) - f(j-1) >}. \tag{14}$$

All the variables appearing in the above expressions, except for λ which is just the magnitude of the first step along the SDM direction, are to be interpreted as vectors. The bracket $< f \mid g >$ notation is just the dot-product of the vectors f and g. The indices i, j or l in parentheses are the iteration numbers and not the component of these vectors. For more detail the reader is referred to References [31] and [32]:

This method has been proven to be very useful in relaxing systems provided the starting point is close enough to the local minimum. There is some freedom with the step size, λ, which needs to be adjusted by trial and error. In addition to optimizations, Broyden's method more generally is used to solve a set of nonlinear equations like $\vec{f}(\vec{x}) = 0$.

Global Optimization Methods [28,29]

A much more challenging task than local optimization methods is to find the global minimum of a multi-valley energy landscape as shown in Figure 1. Global Optimization problems involving a given cost function (minimization of energy or maximization of entropy) arise in many simulation problems dealing with nano systems. This subject has received a great deal of attention in recent years, mostly due to the rapid increase in computer power. The symbolic picture shown in Figure 1 provides a rather simple two-dimensional example of the global optimization paradigm. In this figure the sphere symbolizes a molecule the energy, E(x,y), of which changes as the coordinates (x,y) change. The global optimization objective here is to locate the coordinates (x,y) for which the molecule's enerngy, E(x,y), has its absolute minimum. Of course, this figure shows only a portion of the total energy domain.

For a multi-dimensional space, like phase space, the number of local minima tends to increase quite rapidly and possibly exponentially depending on complexity of the molecule under consideration.

A number of computational algorithms have been developed recently for global optimization of molecular simulations. Among those algorithms the simulated annealing and the genetic algorithms have found more applications in structural and dynamic simulation optimization of molecular clusters. Below, we will describe briefly these two methods.

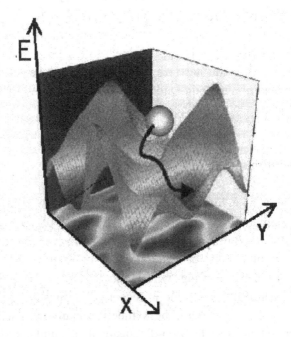

Figure 1. A two-dimensional example of the global optimization of a molecule's energy E(x,y). The objective here is to locate the coordinates (x,y) for which the molecule's energy has its absolute minimum.

1. Simulated Annealing Method: The physical process of annealing is how a metallurgist reaches experimentally the strongest state of a metal sample by following the following three steps:

(i) The metal sample is put in an oven and it is heated to high temperatures.

(ii) At high temperatures the metal atoms are randomly distributed and move about at high speeds.

(iii) Then the sample is cooled very slowly to its original temperature. As a result the atoms settle into crystalline structures, minimizing their energy and rendering the metal much stronger than before.

The implementation of simulated annealing as a global optimization technique was first proposed by Kirkpatrick et al. [33]. This method has attracted significant attention due to its suitability for those large scale optimization problems in which a desired global minimum is hidden among many local minima.

Simulated annealing distinguishes between different local optima and is analogous to the physical process of annealing a metal described above. Namely, one starts the simulation at a sufficiently high temperature. Then the temperature is gradually lowered during the simulation until it reaches the global-minimum-energy state (like physical annealing). The rate of decrease of temperature must be kept sufficiently slow so that thermal equilibrium is maintained throughout the simulation. As a result, only the state with the global energy minimum is obtained. Otherwise, if the temperature decrease is rapid (like hot metal quenching), the simulation will be trapped in a state of local minimum energy.

The statistical mechanical basis of simulated annealing is the Metropolis algorithm presented in the previous chapter. According to the canonical ensemble the probability distribution of a state with energy E_i is expressed by the Boltzmann distribution,

$$p_i \, (T, e_i) = n(e_i). \; \exp(-\beta.e_i), \tag{15}$$

where $\beta = 1/k_B T$ and with the normalization condition,

$$\int p_i \, (T, e_i) d \, e_i = 1. \tag{16}$$

In eq. (15), $n(e_i)$ represents the number of states with energy E_i which is an increasing function of energy.   On the other hand, the Boltzmann factor $\exp(-\beta.E_i)$ is an exponentially decreasing function of E_i.   This gives the probability distribution P_i $(T,\ E_i)$ a well-known bell-shape functional form as shown in Figure 2.

At high temperatures, $\beta=1/k_BT$ is obviously small, and as a result $\exp(-\beta.e_i)$ decreases slowly with e_i . This causes the probability, p_i (T,e_i) to have a wide and short bell-shape. On the other hand, at low temperatures β is large, and $\exp(-\beta.e_i)$ decreases rapidly with e_i. So, p_i (T,e_i) has a narrow, but tall bell-shape (see Figure 2).

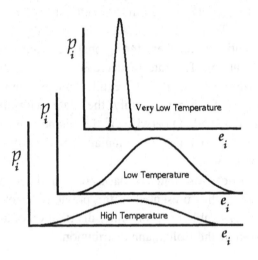

Figure 2. Symbolic variation of the Boltzmann probability distribution function with changes of temperature.

The process of a simulated annealing global optimization of a molecular simulation should possess the following stages:

1. Start the molecular simulation at a rather high temperature to allow the system to rearrange itself from its status quo.

2. Lower the temperature rather slowly to bring the system into a stable (equilibrium) state.

3. Repeat the cycle numerous times so that as many conformations as possible are obtained. Then the global optimum conditions can be found by analysing the results of these conformations.

In simulations it does not matter what dynamics is followed to cross the barriers since the system will be given enough time to come to a stable equilibrium.

The computational code to be written for a simulated annealing should consists of a pair of nested DO-loops. The outer loop varies the temperature according to a "cooling schedule" and the inner loop consist of a Metropolis algorithm at the chosen temperature of the outer loop.

The choice of the cooling schedule is quite important in simulated annealing global optimization. There is a variety of simple cooling schedules to choose from. Amoung them, linear cooling schedule,

$$T_{new} = T_{old} - I(T_o - T_N)/N; \qquad I \leq N, \qquad (17)$$

and algebraic cooling schedule,

$$T_{new} = (I/N) \cdot T_{old}; \qquad I \leq N, \qquad (18)$$

are the most common ones used in practice. Generally, the user may choose an appropriate cooling schedule it by trial and error.

One can also use the adaptive simulated annealing method as proposed by Andersen and Gordon [34] in which the cooling schedule is chosen from the following evolutionary expression for the temperature:

$$dT/d\theta = -vT/\varepsilon\sqrt{C}. \qquad (19)$$

In this equation T is the temperature, θ is time, v is the velocity of annealing (just a constant coefficient controlling the overall cooling rate), C is the heat capacity and ε is the relaxation time,

$$\varepsilon = -Log(\omega_2) \approx 1 - \omega_2, \qquad (20)$$

coming from the second largest eigenvalue of the transition matrix. To construct the transition matrix, one can use a simple model and classify states according to their energy:

1. First, some energy range is defined within which the total energy of the system is fluctuating.
2. This energy interval is then divided into M equal parts (M could be of the order of 100 or even more). Each interval corresponds to one state meaning that if the energy of the system is between E_i and E_{i+1}, we say the system is in the state i ($i=1,M$).
3. Then the system is allowed to evolve at the temperature T according to some dynamics, be it Langevin dynamics or a Markov process defined by the Metropolis algorithm, or even Newtonian dynamics.
4. Then during some time steps P, say of the order of $P=1000$ (this really depends on the size of the system and the complexity of the energy landscape; larger or more complex systems require a larger number of steps), we record the energy of the system, and will increment the matrix element (i,j) of the transition matrix by one unit every time the system goes from state i to state j.
5. If the energy remains in the same interval j, the (j,j) element must of course be incremented by one.
6. At this stage, the elements of each column are rescaled so that their sum is equal to unity. This is the sum rule for the transition matrix as we mentioned in the previous chapter on MC.
7. One now needs to diagonalize this matrix. Its highest eigenvalue must be one, and the corresponding eigenvector contains the equilibrium probabilities for the system to be in each of the M states. From the second eigenvalue one obtains the relaxation time,

$$\varepsilon = -Log(\omega_2) \approx 1 - \omega_2. \tag{21}$$

The heat capacity may be found from the fluctuations in the energy during these P steps run:

$$C = \left[\langle E^2 \rangle - \langle E \rangle^2 \right] / k_B T^2. \tag{22}$$

8. Now that all the elements of the cooling equation are known, one can update the temperature according to: $T = T - vT/\varepsilon\sqrt{C}$ and rerun the dynamics for another P steps, collect the statistics to calculate C and the new transition matrix, and then recalculate the temperature and so on, until it reaches a desired (low) value where the energy fluctuations are below some tolerance level.

2. Genetic Algorithm: Genetic Algorithm (GA) [35,36] is based on concepts coming from genetics and the *"survival of the fittest"* idea; being fit meaning having a lower energy. Genes represent the identity of the individuals. In molecular simulation cases, individuals are clusters and their genes are their coordinates. In GA, individuals are mated among themselves, their genes may also mutate, and this way, one obtains a new generation of individuals selected based on their fitness (potential energy).

The mating, also called the crossover operation, consists of taking two individuals at random, combining half the genes of one with half the genes of the other to obtain a new individual. Note that there are many ways to perform this operation, and one may even produce from a single pair, a next generation of many more individuals which are formed from taking many combinations of the genes of their parents. *A priori*, if we start with N individuals in generation P, one may end up after mating, with say, *20 N* individuals in the next generation. One can eventually perform mutations on them. Mutations consist in randomly modifying part of their genes. In our context, mutation means displacing one or more atoms in a cluster at random. Note that mutation is a local operation, whereas mating or crossover is non-local.

Now from the *20 N* obtained individuals, one can first perform a quick structure optimization, and then sort their energies and pick the N lowest energy individuals as survivors for the next generation. One must, however, be careful to choose candidates of different geometrical forms so as to keep diversity. It would be completely useless to keep N clusters of the same shape for the next generation, even though their energies were the lowest. Now, one can iterate this process and perform mating, mutation, relaxation and final selection based on shape diversity to obtain a new generation and so on... One is hopeful after few hundred or thousand generations (again this depends on the cluster size and complexity of the problem) the fittest individuals will be found. There is however, no guarantee of the convergence of the process.

In SA or GA, one is just sure to have found a good minimum if not the absolute one.

The choice of genes, mating and mutations in GA and the dynamics in SA depend on the problem and can be tailored to the specific problem at hand, just like sampling in Monte Carlo which is not unique to the Metropolis algorithm.

Conclusions and Discussion

The classification of simulation methods presented in this chapter are to give the readers an idea of how every method generally compares with other methods from the points of views of accuracy and CPU.

The importance of the role of molecular dynamics and Monte Carlo computer simulations in nanotechnology is well recognized for several reasons:

1. Such simulation techniques will allow us to develop some fundamental understanding of the behavior of nano systems for which there is presently little or no direct knowledge available.
2. Since in a nano system the number of particles involved is rather small and direct measurement of their collective behavior has not been well developed yet, computer simulations could help appreciably.
3. Computer simulations can produce data for testing and development of analytic predictive models of nano systems. In the next two chapters we introduce Monte Carlo and molecular dynamics simulation techniques for nano systems in detail.
4. For nano systems the main reasons to apply optimization methods is to obtain, either, all the needed local and global low energy conformations of a single molecule or the low energy configurations of an assembly of molecules like a cluster of molecules in an open space or a fluid encapsulated in a nanoscale closed system.

5. An optimization scheme, in general, can give us the opportunity to achieve several important nanotechnology goals: (i). to develop a controlled simulation scheme for obtaining different possible structures by using temperature to surmount torsional barriers of, for example, molecular building blocks; and (ii). to create a computational analog of an experimental procedure to study the most stable conditions of nano systems, such as a self-assembly of molecular building blocks.

Bibliography

[1]. K. Esfarjani and G. A. Mansoori, "*Statistical Mechanical Modeling and its Application to Nanosystems*", *Handbook of Theoretical and Computational Nanoscience and Nanotechnology*, American Sci. Pub., Stevenson Ranch, CA (to appear), (2004).

[2]. M. Foulkes, L. Mitas, R. Needs and G. Rajagopal, *Rev. Mod. Phys.*, **73**, 33-83 (2001); L. Mitas, *Comp. Phys. Commun.* **96**, 107, (1996).

[3]. D. M. Ceperley, *AIP Conf. Proc.* **690**(1) 378 (25 Nov 2003); D. M. Ceperley, *Rev. Mod. Phys* **67**, 279, (1995) and references therein.

[4]. P. Fulde, "*Electron Correlations in Molecules and Solids*", 3rd edition, Springer Series in Sold State, New York, NY, **100**, (1995).

[5]. K. Binder and D.W. Heerman, "*Monte Carlo Methods in Statistical Physics*", 4th edition, Springer (2002); K. Binder, "*The Monte Carlo Method in Condensed Matter Physics*", Vol. **71** in Topics in Applied Physics, Springer-Verlag (1992); K. Binder, *Applications of the Monte Carlo Method in Statistical Physics*", Vol. **36** in Topics in Current Physics, Springer-Verlag, New York, NY, (1984).

[6]. K. D. Sen, "*A celebration of the contribution of R. G. Parr*", Rev. Mod. Chem., World Sci. Pub. Co., Hackensack, NJ, (2002).

[7]. S. R. Langhoff, "*Quantum Mechanical Electronic Structure Calculations with Chemical Accuracy*", Kluwer Acad. Pub., Dordrecht, The Netherlands, (1995).

[8]. L. Hedin and S. Lundqvist, *Solid State Physics* **23**, 1, (1969).

[9]. T.J. Lee and G. E. Scuseria in *Quantum Mechanical Electronic Structure Calculations with Chemical Accuracy*, Edited by S. R. Langhoff, Kluwer Acad. Pub., Dordrecht, The Netherlands, p.47, (1995).

[10]. W. Ekardt, "*Metal Clusters*", John Wiley Series in Theoretical Chemistry, New York, NY, (1999).

[11]. R. G. Parr and W. T. Wang, "*Density Functional Theory of Atoms and Molecules*", Oxford Univ. Press, Oxford, UK, (1989).

[12]. J. R. Chelikowski and S. G. Louie, "*Quantum Theory of Real Materials*", Kluwer Acad. Pub., Dordrecht, The Netherlands, (1996).

[13]. K. Ohno, K. Esfarjani and Y. Kawazoe, "*Computational Materials Science, From Ab initio to Monte Carlo*", Springer series in Solid State, New York, NY, (1999).

[14]. M. C. Payne, M. P. Teter, D. C. Allan, T. A. Arias and J. D. Joannopoulos, *Rev. Mod. Phys.* **64**, 1045, (1993).

[15]. R. Car and M. Parrinello, *Phys. Rev. Lett.* **55**, 2471, (1985).

[16]. M. P. Allen and D. J. Tildesley, *"Computer Simulation in Chemical Physics"*, Nato ASI Series C, Vol. **397**, Kluwer Acad. Publ., Dordrecht, The Netherlands, (1992).

[17]. P. E. A. Turchi, A. Gonis and L. Colombo, *Tight-Binding Approach to Computational Materials Science*, MRS Proceedings (1998); P. E. A. Turchi and A. Gonis, *Electronic Structure and Alloy Phase Stability*, JPCM **13**, 8539-8723, (2001).

[18]. W. A. Harrison, *"Elementary Electronic Structure"*, World Sci. Pub. Co., Hackensack, NJ, (1999).

[19]. M. Sprik in *"Computer Simulation in Chemical Physics"*, Nato ASI Series C, Vol. **397**, Kluwer Acad. Pub., , Dordrecht, The Netherlands, p.211 (1992).

[20]. J. Israelachvili, *"Intermolecular and Surface Forces"*, 2nd Edition, Academic Press, New York, NY, (1995).

[21]. M. E. J. Newman and G. T. Barkema, *"Monte Carlo Methods in Statistical Physics"*, Clarendon Press, Oxford, UK (1999).

[22]. D. P. Landau, K. K. Mon and H. B. Schuttler, *"Computer Simulation in Condensed Matter Physics IV"*, Springer, Berlin, Germany, (1992).

[23]. H. Gould and J. Tobochnik, *"An Introduction to Computer Simulation Methods"*, 2nd edition, Addison-Wesley, Boston, MA, (1996).

[24]. J. Singh, *"Electronic and Optoelectronic Properties of Semiconductor Structures"*, Cambridge Univ. Press, Cambridge, UK, (2003).

[25]. W. Krauth and O. Pluchery, *J. Phys. A*: Math Gen **27**, L715 (1994); W. Krauth, cond-mat/9612186 *"Introduction To Monte Carlo Algorithms"* in *Advances in Computer Simulation*, J. Kertesz and I. Kondor (Ed's), Lecture Notes in Physics (Springer Verlag, 1998); W. Krauth, cond-mat/0311623 *"Cluster Monte Carlo algorithms"* Chapter of *New Optimization Algorithms in Physics*, A. K. Hartmann and H. Rieger (Ed's), (Wiley-VCh) (2004); J. M. Pomeroy, J. Jacobsen, C. C. Hill, B. H. Cooper and J. P. Sethna, *Phys. Rev. B* **66**, 235412 (2002).

[26]. K. K. Bhattacharya and J. P. Sethna, *Phys. Rev. E* **57**, 2553 (1998).

[27]. R. Fletcher, *"Practical Methods of Optimization"*, 2nd Edition, John Wiley & Sons, New York, NY, (1987).

[28]. C. A. Floudas and P.M. Pardalos, eds., *"Recent Advances in Global Optimization"*, Princeton Series in Computer Science, Princeton Univ. Press, Princeton, NJ, (1991).

[29]. M. Lipton and W.C. Still, *J. Comput. Chem.*, **9**, 343, (1988).

[30]. W. H. Press, S. A. Teukolsky, W. T. Vetterling and B. R. Flannery; *"Numerical Recipes, The Art of Parallel Scientific Computing"*, Cambridge Univ. Press, Cambridge, UK, (1996).

[31]. C. G. Broyden, *Math. Comput.* **19**, 577, (1965).

[32]. D. Singh, *Phys. Rev. B* 40, 4528 (1989).

[33]. K. S. Kirkpatrick, C. D. Gelatt and M. P. Vecchi, *Science* **220**, 671, (1983).

[34]. B. Andersen and J. M. Gordon, *Phys. Rev. E* **50**, 4346, (1994).

[35]. D. E. Goldberg; *"Genetic Algorithms in Search, Optimization and Machine Learning"*, Addison-Wesley, Boston, MA, (1989).

[36]. D. A. Coley, *"Introduction to Genetic Algorithms for Scientists and Engineers"*, World Sci. Pub. Co., Hackensack, NJ, (1999).

Chapter 7

Phase Transitions in Nanosystems

"We may call such bodies as differ in composition or state, different phases of the matter considered, regarding all bodies which differ only in quantity and form as different examples of the same phase." **J. Willard Gibbs**

Introduction

In this chapter the historical perspective of phase transitions is presented including the principles of phase transitions in macroscopic systems due to Gibbs. The phenomenon of fragmentation which has been observed in small system phase transitions is introduced. Comparisons between the first order phase transitions in small, intermediate and large scale systems are made based on the concept introduced by S.A. Berry. A number of experimental examples of phase transitions in small systems are introduced. It is argued that a better understanding of phase transitions in small systems will help the development of self-assembly which is the fundamental of bottom up nanotechnology.

Principles of phase separations / transitions are well-defined and formulated in the macroscopic limit [1-6]. The macroscopic limit is defined when the number of atoms and molecules in a large system is of the order of 10^{23} or more which is theoretically at infinite limit, known as thermodynamic limit [(N and V)$\to\infty$ but N/V=finite]. Most of the existing theories of fluids, solids, mixtures and phase transitions are developed for systems at the thermodynamic limit using, mainly, canonical or grand-canonical ensembles [7,8]. However, for small systems consisting of limited number of particles, principles governing

191

the separation of phases is not well understood yet. Actually the thermodynamic property relations for small systems are also not clear and open for further study. The well known thermodynamic property relations, such as the van der Waals, perturbation and variational theories of equation of state [4, 9, 10], etc., are defined and valid only for extensive large systems in the thermodynamic limit [12].

Investigation into phase separations, phase transitions and phase changes in macroscopic systems, both, experimentally and through the applications of statistical mechanics, has a long tradition going back to 1800s. Andrews was probably the first person who studied the first order phase transition (with phase separation due to density difference and interfacial tension between phases) in 1869 ending the earlier notion of "permanent gases" [11]. Before Andrews there was this ungrounded concept of "permanent gases" due to the fact that one could not condense a gas, like air, by simply compressing it without lowering its temperature. At that time the concept of critical point was not yet recognized and the understanding that a gas needs to go below its critical temperature to become condensable to a liquid was not discovered yet. J.D. van der Waals in 1873 worked on, probably, the first recorded predictive theory of phase transitions which resulted in the well recognized van der Waals equation of state (vdW eos) [11],

$$(P + \frac{a}{v^2})(v - b) = RT . \tag{1}$$

This equation of state, when solved for pressure versus volume, produced all the necessary fluid phases which was experimentally observed by Andrews that could co-exist and gave a simple and, at that time, satisfactory account of phase transitions. Van der Waals was, probably the first scientist who proved that phases are changed between liquids and gases. The vdW eos is now modified in various forms and its modifications have found many applications in the analysis of thermodynamic properties of pure fluids and mixtures in macroscopic systems [5,11,13,14] of scientific and industrial interest. It's theoretically based extensions and modifications have been the subject of many theories of statistical mechanics [1,6,7,11]. The basic assumption in deriving the vdW eos and majority of the other existing statistical mechanical theories of fluids and solids, is that the dimensions of the macroscopic system are very large when compared with those of the constituting atoms and molecules.

The Gibbs Phase Rule

After the discoveries of van der Waals, investigations into phase transitions in macroscopic systems consisting of mixtures of many components in vapors, liquids and solid phases became a major subject for research in late 1800 until the early 1900. Among all such investigations, in line with phase transitions, one should mention the well recognized research findings of J. Willard Gibbs [15] in late 1800. Gibbs gave the phase transitions in systems at equilibrium a firm theoretical ground and formulated his famous "phase rule" which today is well known in all fields of science and engineering. The Phase Rule describes the possible number of degrees of freedom in a (closed) macroscopic system in thermodynamic limit at equilibrium. The general form of the phase rule for a multiphase-multicomponent-reacting system is:

$$F = c - p - r + 2. \tag{2}$$

In formulating the phase rule, the **Degrees of Freedom**, F, is defined by Gibbs as the number of independent intensive variables (i.e. those that are independent of the quantity of material present) that need to be specified in value to fully determine the state of a macroscopic system.

A **Phase**, as defined in a macroscopic system, is a component part of the system that is immiscible with the other parts (e.g. solid, liquid, or gas); a phase may, of course, contain several chemical constituents, which may or may not be shared with other phases. The number of phases is represented in the phase rule relation by p.

The **Chemical Constituents** are simply the distinct compounds (or elements) present in the system. When some of the system constituents remain in reaction equilibrium with each other whatever the state of the system, they are counted as a single constituent and every independent chemical reaction will be considered as a constraint, reducing the degrees of freedom of the system. The number of components is represented as c and the number of independent chemical reactions is represented as r.

For example, a macroscopic system with one component and one phase (a gas cylinder full of helium, perhaps) has two degrees of freedom: temperature and pressure, say, can be varied independently. Of course, so long as there is one component in the system there can't be any chemical reactions and $r=0$. However, for a macroscopic system with two reacting components and one phase (a gas cylinder full of ozone, O_3, and oxygen, O_2, for instance) where the equilibrium reaction $2O_3 \leftrightarrow 3O_2$ is present it again has two degrees of freedom. For a two phase pure system, a liquid and vapor for instance, you lose a degree of freedom, and there is only one possible pressure for each temperature. Add yet one more component with no reaction, then you have two degrees of freedom. For a one component and three phases $F=0$, and so is for a two component and two phase with a chemical reaction.

In the case of small / nano systems due to the lack of experimental measurement techniques we have not been able to study phase transitions directly. However, investigations either through indirect experimental methods or by computer simulation techniques have been successful to distinguish various phases of matter in small systems, as small as clusters of a few atoms or molecules [16-18]. Nano, or small, systems are those where the linear dimension of the system is of the characteristic range of the interaction between the particles comprising the system. Of course, astrophysical systems are also in this category and from the point of view of interactions between planets and stars they may be considered "small". Even though the principles of phase transitions are not well defined for small systems, there are many phenomena in small systems that resemble the phase transitions in large systems. This has been specially the case in the study of clusters of tens, hundreds and thousands of molecules by various investigators [17,18].

Considering that nanoscale systems consist of finite number of particles (intermediate in size between isolated atoms / molecules and bulk materials), principles of phase transitions, as formulated for large systems, need to be reformulated for nano systems [12,17,18]. While we can have control-mass (closed) nano systems, however due to the fact that such systems are not in thermodynamic limit and they may not be in equilibrium, from the point of view of Gibbsian thermodynamics, we may not be able to use the Gibbs phase rule for such systems. Also, for nano systems, which are actually nonextensive, the definition and separation of extensive and intensive variables is not quite clear, and very possibly, they depend on the size of the system.

Phase Transitions

Principles of phase separations / transitions are well defined and formulated in the thermodynamic limit [$(N\&V) \rightarrow \infty$ N/V=finite] as it is demonstrated for the PVT relation of pure systems depicted by Figures 1 and 2.

In Figure 1, the pressure versus volume projection of a pure large system is presented and various isotherms, phases of matter including solid (S), liquid (L) gas (G), ideal gas (IG), supercritical fluid (SCF) and vapor (V) are distinguished. In Figure 2, the pressure versus temperature projection of the same system is presented and various phases of matter are separated by phase transition curves. Also reported in these figures are the critical point (CP), triple point (TP), and various phase transition regions. There exist two kinds of phase transitions in large systems as is

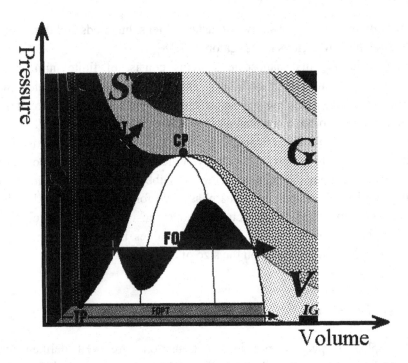

Figure 1. Pressure vs. volume projection of a simple pure substance phase transitions/ equilibrium in the thermodynamic limit.

demonstrated in Figure 1. The first order phase transitions (*FOPT*) and the second order phase transitions (*SOPT*). The distinct feature of *FOPT* is the existence of density and surface tension difference between the two phases which reveals itself in the form of meniscus. While in the *SOPT* such differences between the phases do not exist. *FOPT* are distinguished from continuous transitions, *SOPT*, by the appearance of phase-separations. In *FOPT* the system becomes inhomogeneous and coexistent phases are separated by interfaces. Of course, a system defined at phase separation is necessarily inhomogeneous, not satisfying the thermodynamic limit and may be considered non-extensive.

Thermodynamic property relations in nano systems are functions of the geometry and internal structure of the system under consideration. In contrast to thermodynamics of large systems, the thermodynamic properties and property relations of small systems will be generally different in different "environments". As a result appropriate ensemble technique must be used for different nano / small systems in order to predict their behavior [12].

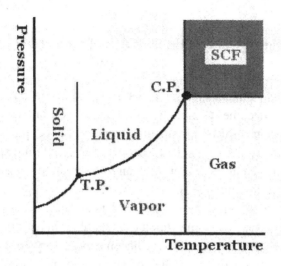

Figure 2. Pressure vs. temperature projection of a simple pure substance phase transitions/ equilibrium in the thermodynamic limit.

Understanding of phase transition in systems composed of finite number of particles is a peculiar and unsolved problem from, both, the theoretical and experimental points of view. Considering that nanoscale systems consist of finite number of particles (intermediate in size between isolated atoms / molecules and bulk materials) principles of phase transitions need to be reformulated for nano systems.

Nano systems composed of limited number of particles and confined to nanoscale volumes are not large enough compared to the range of the interatomic and intermolecular forces existing between their particles.

A more precise characterization of all such small systems, is to call them nonextensive systems as discussed in Chapter 3. According to Tsallis [19], the test of nonextensivity of a system is in the non-additivity of its entropy. In another words, if a nonextensive system is divided into pieces, the sum of the entropies of its parts will not be equal to the entropy of the original system.

A Comparison of Phase Transitions Between Small and Large Systems

Having a good understanding of phase transitions in large systems, a question always comes to mind on how the very few and recent observed phase transition effects in small systems, can be related to well-understood phase transitions in macroscopic systems (in thermodynamic limit). Experimental measurements and observations regarding small systems are quite limited [12].

In order to develop some basic understandings about nano system phase transitions, and due to the difficulty in performing direct experimental measurement, it has become necessary to use computer simulation methods of assemblies of atoms and molecules. One of the useful and revealing computational techniques regarding the study of phase transitions in small systems is the utilization of such techniques as Monte Carlo (MC) and Molecular Dynamics (MD) approaches presented in the previous chapters. There have been a number of such computational studies performed in recent years by various research groups from different disciplines of science and engineering [17,20-22].

There have been a great deal of research progress in the development of understanding the phase transitions of a class of small systems which has been known as molecular clusters consisting of a few (tens or hundreds) of atoms or molecules. Most such studies are performed using MC and MD techniques [19,20-22]. Such computations have helped us to understand some of the possibilities of phase transitions in small systems and the relationship of such transitions to large systems. They also have revealed some unique properties of phases and phase transitions in nanoscale systems, which are not known, to large systems.

A very useful study resulting from phase transition computer simulations and comparison of such results for small and large systems is a comprehensive study made by R.S. Berry and coworkers [17,18]. In order to compare phase transitions in small and large systems these investigators used the following criteria, which are in common between all systems whether small, intermediate or large: They proposed that the essential condition to be considered for *local* stability (as well as metastability) of any form of matter, is the occurrence of a local minimum in the free energy, for fixed external variables, usually intensive, with respect to some suitable order parameter [17,18]. They also considered the equilibrium constant $K(T) \equiv [L]/[S]$. In this expression $[L]$ represents the amount of liquid phase and $[S]$ represents the amount of solid phase. This essential relationship determines the relative proportion of two locally stable phases (for example a liquid and a solid phase) with different free energies. Similar to the principle of chemical equilibrium, one may assume the following general expression will hold for the equilibrium constant, $K(T)$ of two phases in contact with respect to their free energy difference [17,18]:

$$K(T) = \frac{[L]}{[S]} = \exp\left(-\frac{\Delta G}{Nk_B T}\right), \qquad (3)$$

where $\Delta G(T)$ is the difference in free energies of phases L and S at temperature T provided other system variables are kept constant and uniform between the phases. The free energy difference can then be written in terms of the mean chemical potential difference $\Delta\mu(T)$ between the two phases and the number of particles, N, in the system as the following [18],

$$\Delta G(T) = N\Delta\mu(T). \qquad (4)$$

At local minimum in the free energy we will have $\Delta G=0$ which, according to Eq. (4) is equivalent to $\Delta\mu=0$. This condition is now in line with the Gibbsian thermodynamics which states that two phases can coexist in equilibrium only when $\Delta\mu = 0$. It is important to note that this condition will be valid regardless of the size of the system under consideration, whether in nanoscale consisting of a countable number of particles (atoms or molecules) or in macroscopic scale with 10^{23} or more particles.

Berry and coworkers [17,18] also conclude that one can expect to observe both phases in the nanoscale ensemble over a range of temperatures and not just at a single temperature as it is the case for macroscopic systems. The range of conditions, in the occurrence of a local minimum in the free energy for the phases, sets the limits within which coexistence of phases is possible.

As an example the liquid-solid phase transition for three distinct systems, one small consisting of a few particles, one of intermediate size and one large (macroscopic) in the thermodynamic limit were studied by Berry and co-workers and they are illustrated schematically in Figures 3-5.

Small System

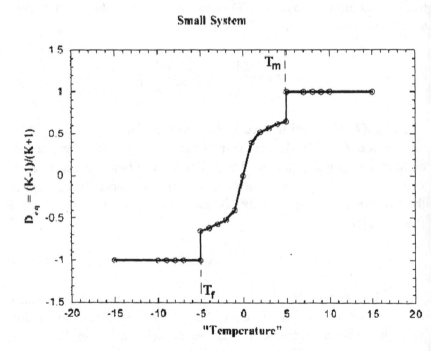

Figure 3. The coexistence of two phases of a quite small system. Two distinct transition points, the freezing, T_f, and the melting, T_m, points are recognized for this system. Large discontinuities at T_f and T_m are observed [19].

Mid-Size System

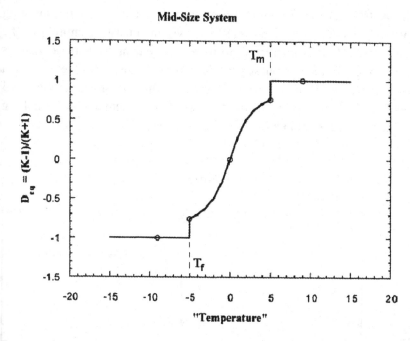

Figure 4. The coexistence of two phases of an intermediate size system. Still two distinct transition points, the freezing, T_f, and the melting, T_m, points are recognized for this system like in Figure 3. Large, but less pronounced discontinuities at T_f and T_m are still observed [19].

The two-phase region of the macroscopic system, Figure 5, is slightly modified by us to make it conform to experimental macroscopic systems. In these figures the vertical coordinate is the distribution parameter D defined as,

$$D = \frac{K-1}{K+1} = \frac{[L]-[S]}{[L]+[S]} = \frac{e^{-\Delta\mu/k_B T}-1}{e^{-\Delta\mu/k_B T}+1},\qquad(5)$$

and the horizontal coordinate represents a dimensionless temperature normalized to be zero at the middle of the transition region. With this choice of the vertical coordinate, then D obviously is the amount of liquid minus the amount of solid, divided by the total amount of material.

The distribution parameter D varies from (-1) for a system consisting only of the solid phase S to (+1) for a system consisting entirely of the liquid phase L. It is interesting to note that systems in thermodynamics limit posses only one phase transition temperature. However, small systems, which are not in the thermodynamic limit as shown in Figures 3 and 4, always possess two distinct phase transition temperatures (melting T_m and freezing T_f points for solid-liquid transition).

Large System

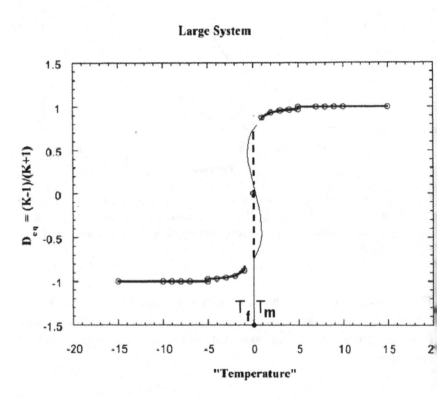

Figure 5. The coexistence of two phases of a large system (in the thermodynamic limit) [19]. The two distinct transition points (freezing T_f and melting T_m) which were recognized in small- and intermediate-size systems (Figures 3 and 4) have reduced to one point in this case and the discontinuities between the two phases have disappeared. The two-phase region is slightly modified by us to make it conform to experimental macroscopic systems.

In these figures the region of the graph corresponding to $D = -1$ represents the solid phase, the region corresponding to $D = +1$ represents the liquid phase. The transitions region between the freezing point, T_f, and melting point, T_m, is quite different between small, intermediate and large systems. As is shown in Figures 3 and 4 discontinuities are observed at the transition temperatures, T_f and T_m, of small and intermediate size systems and such discontinuities become more pronounced as the size of the system decreases. For large systems the freezing and melting points coincide and the discontinuity in the two phase region disappears as it is demonstrated in Figure 5. This behavior is associated with the first-order phase transitions as it is well recognized in large systems.

The above discussion and the graphical representations are, all, for the first-order phase transitions of systems of various sizes. Another question of importance is about the nature of the second-order phase transitions in small systems. In the first order phase transitions in macroscopic systems we observe abrupt changes in entropy and energy associated with the phases, and physically, there is a distinct phase separation (meniscus in the case of liquid and vapor transition) apparent between the phases. The second order phase transitions, on the other hand, do not exhibit any abrupt change. But there are abrupt changes, discontinuities, in the derivatives of the energy with respect to, for example, temperature such as the heat capacities. While some studies have been made on special cases of the second-order phase transitions in small systems, the details of such problem are still under investigation [18].

Fragmentation

Fragmentation is considered a ubiquitous phenomenon by many investigators, and has been the focus of a great deal of study in recent years. Fragmentation underlies processes such as polymer degradation, breakup of liquid droplets, crushing of rocks and breakup of continuous phases, in particular in small systems, during phase transitions. Fragmentation was originally observes in nuclear systems, now known as nuclear multifragmentation. Fragmentation as a real phase transition of first order in nuclei was theoretically interpreted and computationally simulated by several investigators [23-29].

Fragmentation is a new generic structural transition of interacting finite many-body systems. Because of fragmentation, finite systems have a new degree of freedom unknown to systems in thermodynamic limit. In addition to the well-known structural phase transitions as melting or boiling, fragmentation into several large fragments and possibly also into many monomers or small clusters is characteristic in the disintegration of finite / small systems. It is similar to boiling in macro systems but has many new features due to the important size fluctuations, which distinguishes fragmentation from the simple phase separations.

For example, comparing the boiling phenomena in macro and nano systems [28] it is observed that a liquid in a nano system also goes through fragmentation before evaporation. This is due to the important size fluctuations and correlations in small system liquids as their temperature is increased. Fragmentation distinguishes a phase transition in small systems from the same kind of transition in large systems.

Fragmentation of fluids occur when the ratio of viscous to capillary forces, exceeds a critical value which depends on the interfacial tension, the viscosity ratio between dispersed and continuous phases, pressure, temperature and the nature of the flow, for systems in flow condition [26].

A finite, small system, at phase transition of first order, fragments into various coexisting regions of different phases, e.g. liquid and gas. In general, the system loses its homogeneity and develops global density fluctuations. In macroscopic systems where conventional thermodynamics of coexistence of phases is valid, phase transition is usually treated by considering two infinite homogeneous regions separated, for example, by an external gravity field. Then the two phases exchange energy and particles through an interfacial surface.

The formulation and prediction of fragmentation as a real phase transition of first order in a "Small" many-body system has been a challenge in statistical mechanics to understand its foundation better and also to be able to describe the thermodynamics of inhomogeneous and nonextensive systems. It opens thermo-statistics for so many applications from small systems like nuclei up to the largest like astrophysical ones.

Experimental Observations of Phase Transitions in Small Systems

1. Evaporation of Water in a Sealed Nanotube

A recent experimental observation of phase transition in nano systems deals with the evaporation of water inside a sealed multi-wall carbon nanotube [30]. It is shown that multi-wall carbon nanotubes are quite impenetrable even when heated under the high vacuum (1028 Torr) of the transmission electron microscope (TEM) Column. Because of this, multi-wall carbon nanotubes could be used as arteries for mass transfer in nanotechnology, similar to regular pipelines used in the macroscopic world. Typically, the inner diameter of a multi-wall carbon nanotube is up to 10 ϕ [nm] and the wall thickness could exceed the inner diameter.

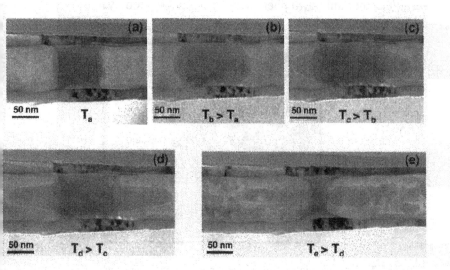

Figure 6. First order phase transition of water inside a multiwall carbon nanotube as photographed in a TEM [31]. We recognize fragmentation of the liquid phase as the temperature is increased.

The sealed multi-wall carbon nanotube containing water (and possibly some CO_2 and CH_4 impurities) was synthesized hydrothermally using Tuttle-type tube autoclaves at pressures of up to 100 MPa and at temperature range of 700–800 °C in the presence of nickel catalyst [31]. Some of the carbon nanotubes produced by this method were found to contain segregated liquid and gas phases with well-defined interface observable clearly under a transmission electron microscope (TEM). The pressure inside the nanotube at room temperature was estimated to be up to 30 MPa. High pressures are known to facilitate increased solubility of CO_2 and CH_4 gases in water and possibility enhancement of wettability of water on the inside-wall of carbon nanotube.

In Figure 6 the electron microscope image of a peculiar behavior of water in a nanotube under variation of temperature is shown as reported in Ref. [30]. This figure demonstrates the first order phase transition, evaporation of water, in a multiwall carbon nanotube using TEM. Fragmentation of the liquid phase during evaporation is well demonstrated in this figure as temperature is raised through electron beam. Whether the observed fragmentations are due to interfacial effects, and/or other unknown phenomena, is not understood yet.

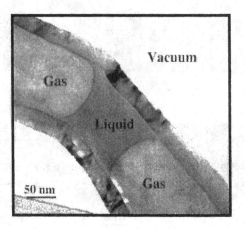

Figure 7. Demonstration of the phase separation and fragmentation in the first order phase transition of water inside a mutiwall carbon nanotube. These figures correspond to the same experiment as shown by Figure 6 as photographed in a TEM [31,34].

Another recent observation of the behavior of fluid water in a nanotube [31-33] is what is shown in Figure 7. According to this figure the vacuum condition outside of the multiwall carbon nanotube is not removing the water molecules from the nanotube. This is an indication that the multiwall carbon nanotube is impenetrable by water molecules.

2. Micellization and Coacervation

A micelle consists of a reversible assembly of molecules such as surfactants that assemble together in a solution. Micellization is a phenomenon originally observed due to the self-association process of the surface active materials in aqueous solution. These surface active materials, known as surfactants, tend to self-assemble into geometric (disks, spheres, cylinders) shapes and become suspended in the solution. This phase transition phenomenon occurs only when the surfactant concentration exceeds a threshold known as critical micelle concentration (CMC). The change in properties that occur as micelles form is marked by sharp transitions in many physical quantities such as the surface tension, viscosity, conductivity, turbidity and nuclear magnetic resonance of the solution [34]. Although micellization represents a self-association phenomenon initiated by the hydrophobic-hydrophilic imbalance it is a phase transition in nanoscale. The hydrophobic part of the surfactant molecule tends to avoid contact with water, while the hydrophilic ionic head group tends to be strongly hydrated. Self-association of surfactant molecules into micelles can be seen as resulting from a compromise between the two different properties of the surfactant molecules [35].

If one starts with surfactant molecules in an appropriate medium, at the CMC, surfactant molecules change phase and enter in the micelle (nano) phase. Then the micelles in an appropriate medium (like in liquid state), and as a result of increase in their concentration, again change phase into a new phase. The original micelle phase system then becomes two phases. One is the micelle phase and the other being the micelle-coacervate phase. The micelle-coacervate phase is in a dispersed state and it could appear like liquid droplets. Upon standing, these coalesce into one clear homogenous micelle-rich liquid layer, known as the coacervate layer which can separate to a new phase [36,37]. The Coacervation process was originally discovered and developed into practical applications by Green and Schleicher [38]. Coacervation phase separation from micelles involve many properties, materials and processes such as phase inducing agents, stirring rates, core to wall ratios, amphiphile polymer characteristics, core characteristics (wettability, solubility), cooling rates etc. [36,37].

MOLECULES ⟺ MICELLES ⟺ MICELLE-COACERVATES

In Figure 8 the phase transition phenomena in the process of micellization of an asphaltene amphiphile is reported [36,37].

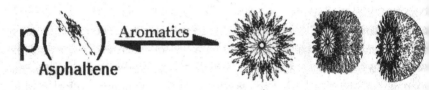

Figure 8. Micellization phase transition (self-assembly) of asphaltene amphiphiles (surfactant molecules) due to an increase in polarity (aromaticity in this case) of its medium [37,38].

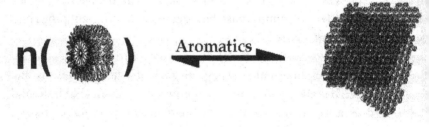

Figure 9. Coacervation phase transition (self-assembly) of asphaltene micelles due to an increase of micelle concentration and in an increasing polarity (aromaticity) medium [37,38].

Well-characterized asphaltene amphiphiles separated from petroleum could constitute a molecular building block in nanotechnology. Also in Figure 9 the phenomena of coacervation phase transition of asphaltene micelles [36,37] are reported. Coacervation of miclles is a potential self-assembly process which is being considered in nanotechnology. It should be pointed out that the molecular theories of micellization and coacervation [39,40] are well developed and can be utilized in practical application of such self-assembly processes.

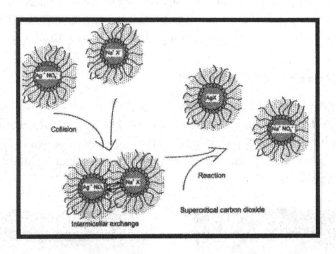

Figure 10. The proposed mechanism for the formation of AgI nano particles (micelle coacervates) in supercritical CO_2 by mixing two water-in-CO_2 microemulsions with one containing $AgNO_3$ and the other NaI in the water core [49].

Instances of micellization in various media (supercritical fluid and liquid phases) consist of polymerizable amphiphiles, polymer and lipid utilization to form ordered aggregates like mono/bilayers, micelles and nanoparticles. Recently a number of techniques for synthesis of well-dispersed micelles (known as emulsions) within porous solid structures were accomplished [41,42]. These new techniques are expected to have an important impact into the development of more efficient nanocatalysts. This was done by using supercritical carbon dioxide as the deposition and synthesis medium [43]. Supercritical fluids and their applications in separations, extraction and synthesis is an active field of research and applications [44-47]. The nanoscale regions in the micellar structures could also act as templates for manufacturing of semiconductor and metalic nanoparticles. In these systems, the contents of different micellar spaces would undergo exchange when the micelles collide, creating opportunities for mixing and reactions between the reactants in different cavities [43].

By using microemulsions carrying different ionic species in the water cores [48] silver halide nanoparticles (micelle coacervates) were produced in supercritical CO_2. Figure 10 shows the dynamics of the two microemulsions explaining formation of the AgI nanoparticles.

3. An Example of Crystallization

Liquid to solid phase transition and crystal formation is one of the major areas of interest in nanotechnology. Nanocrystals are in the category of building blocks of nanotechnology. There have been several processes developed for the production of semiconductor, metallic and organic nanocrystals.

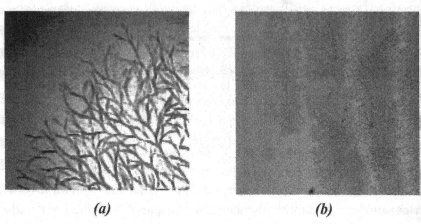

(a)   *(b)*

Figure 11. An atomic force microscope image of banded growth of (a) natural wax and (b) wax mixed with a crystallization inhibitor. A route for nano crystal formation in liquid to solid phase transition [Courtesy of Prof. J.L. Hutter].

One example for the study of conditions for formation and production of nanocrystals is depicted in Figure 11. In this study an atomic force microscope is used to visually study crystallization of liquid wax when the temperature is lowered. Varying the temperature gradient causes a transition between the growth of wax plates and growth of a tree-like structure with regular branches. This figure illustrates two growth images of the wax "tree" due to lowering the temperature and causing solidification of wax and its separation from solution. In the left image natural wax is crystallized. While in the right image wax mixed with a crystallization inhibitor is crystallized. While the intention of the investigators have been to inhibit crystallization, it is clear that another role of inhibitor is production of wax nanocrystals. Understanding the molecular details of wax crystallization has been of major industrial interest in transportation of waxy petroleum crude oils as well as the applications of wax [49,50].

Figure 12. An AFM image of the spiral growth of paraffin crystal (measuring approximately 15 microns across) [Courtesy of Prof. M.J. Miles].

Figure 12 shows an atomic force microscope image of the spiral growth of paraffin crystal (measuring approximately 15 microns across). Inset shows orthorhombic arrangement $(0.49\phi\ [nm] \times 0.84\phi\ [nm])$ of chain ends of one of the crystal terraces.

It is presently difficult, if not impossible, to come up with universal thermodynamic property relations valid for all nano systems [12]. However, computer simulation techniques as discussed later in this report may allow us to predict properties and phase transitions behavior of small systems. In order to apply statistical mechanics to nano systems it is also necessary to measure or compute the intermolecular interactions in nano systems, which is the subject of the next section of this report.

Conclusions and Discussion

Due to lack of thermodynamic limit and appreciable fluctuations in temperature and pressure in small systems, as compared to macroscopic systems, the following peculiarities in small systems are expected to happen during phase transitions:

1. Phase transition in nanoscale systems is, generally, a dynamic phenomena and not of the static (equilibrium) nature as it is in macroscopic phase transition which is called phase equilibrium.
2. Coexistence of phases in small systems are expected to occur over bands of temperatures and pressures, rather than only at a sharp point as is the case for large systems.
3. The Gibbs phase rule loses its meaning; and many phase-like forms may occur for nanoscale systems that are unobservable in the macroscopic counterparts of those systems.
4. Like large systems, various forms of phase transitions, including the first order, second order, and more, are expected to occur in small systems.

5. Small systems have an additional degree of freedom unknown to large (infinite) systems. In addition to the well-known structural phase transitions as melting or boiling, fragmentation into several large fragments and possibly also into many monomers or small clusters is characteristic for the disintegration of finite systems. It is similar to boiling in macro systems but has many new features due to the important size fluctuations and correlations which distinguish fragmentation from the more simple evaporation.

6. It should be pointed out that the science of phase transitions is the basic science behind molecular self-replication.

Bibliography

[1]. K.H. Hoffman and Q. Tang, *"Ginzburg-Landau Phase Transition Theory and Superconductivity"*, ISNM **134**, 383, Birkhauser-Verlag, Basel, Germany, (2001).

[2]. D. Kinderlehrer, R.D. James and M. Luskin (Ed's), *"Microstructure and phase transition"*, IMA volumes in mathematics and its applications, Springer-Verlag, 1-25, (1993).

[3]. E. Z. Hamad and G. A. Mansoori, *J. Phys. Chem.*, **94**, 3148, (1990).

[4]. S. S. Lan and G. A. Mansoori, *Int'l J. Eng. Sci.*, **14**, 307, (1975); *ibid*, **15**, 323, (1977).

[5]. G. A. Mansoori and J. F. Ely, *J. Chem. Phys.*, **82**, 406, (1985).

[6]. P. Pourghesar, G. A. Mansoori and H. Modarress, *J. Chem. Phys.*, **105**, 21, 9580, (1996).

[7]. J. M. Haile and G. A. Mansoori, (Ed's) *"Molecular-Based Study of Fluids"*, Avd. Chem. Series **204**, ACS, Wash., D.C., (1983).

[8]. E. Matteoli and G. A. Mansoori (Ed's), *"Fluctuation Theory of Mixtures"* Taylor & Francis, New York, NY, (1990).

[9]. G. A. Mansoori and F. B. Canfield, *J. Chem. Phys.*, **51**, 11, 4967, (1969).

[10]. A. H. Alem and G. A. Mansoori, *AIChE J.*, 30, 475, (1984).

[11]. G. A. Mansoori and J. M. Haile, *Avd. Chem. Series*, **204**, 1, ACS, Wash., D.C., (1983).

[12]. G. R. Vakili-Nezhaad and G. A. Mansoori, *J. Comp'l & Theo'l Nanoscience*, **1**, 233, (2004).

[13]. T. Y. Kwak, E. H. Benmekki and G. A. Mansoori, in *ACS Symp. Series*, **329**, 101, ACS, Wash., D.C., (1987).

[14]. A. P. Pires, R. S. Mohamed and G. A. Mansoori, *J. Petrol. Sci. & Eng.*, **32**, 103, (2001).

[15]. J. W. Gibbs, *"The Collected Works of J. Willard Gibbs"* **I** & **II**, Yale Univ. Press, New Havens, Conn., (1948).

[16]. T. L. Hill, *"Thermodynamics of Small Systems"*, Parts I and II, Reprint Edition Dover Pub., City, State, (1994).

[17]. R. S. Berry, in: *"Theory of Atomic and Molecular Clusters"*, J. Jellinek (Ed.), Springer-Verlag, Berlin, Germany, (1999).

[18]. R. S. Berry, *C. R. Physique*, **3**, 319, (2002).

[19]. C. Tsallis, R. S. Mendes and A. R. Plastino, *Physica A* **261**, 534, (1998).

[20]. J. S. Kaelberer and R. D. Etters, *J. Chem. Phys*. **66**, 7, 3233, (1977).

[21]. O. Mulken, H. Stamerjohanns, and P. Borrmann, *Phys. Rev. E*, **64**, 047105, (2002).

[22]. N. A. Alves and J. P. N. Ferrite, *Phys. Rev. E*, **65**, 036110, (2002).

[23]. Z. Cheng and S. Render, *Phys Rev Lett.*, **60**, 24, 2450, (1988).

[24]. X. Campi, *Phys.Lett.B*, **208 B**, 351, (1988).

[25]. D. H. E. Gross, *"Microcanonical Thermodynamics"*, World Scientific Lecture Notes in Physics **66**, World Sci. Press, Singapore, (2001).

[26]. J. P. Bondorf, A. S. Botvina, A. S. Iljinov, I. N. Mishustin, and K. Sneppen, *Phys.Report*, **257**, 133, (1995).

[27]. L. Courbin, W. Engl, and P. Panizza, *Phys. Rev. E* **69**, 061508, (2004).

[28]. G.S. Fanourgakis, P. Parneix and Ph. Bréchignac, *Eur. Phys. J. D* **24**, 207, (2003).

[29]. A. J. Cole, *Phys. Rev. C* **69**, 054613, (2004).

[30]. Y. Gogotsia, J. A. Liberab, A. Guvenc-Yazicioglu, and C.M. Megaridis, *Applied Phys. Lett.*, **79**, 7, 1021, (2001).

[31]. J. Libera and Y. Gogotsi, *Carbon*, **39**, 1307, (2001); Y. G. Gogotsi, J. Libera, and M. Yoshimura, *J. Mater. Res.*, **15**, 2591, (2000).

[32]. C. M. Megaridis, A. G. Yazicioglu and J. A. Libera, *Physics of Fluids*, **14**, 2, 15, (2002).

[33]. Y. G. Gogotsi, T. Miletich, M. Gardner and M. Rosenberg, *Rev. Sci. Instrum.*, **70**, 12, 4612, (1999).

[34]. S. I. Stupp, V. LeBonheur, K. Wlaker, L. S. Li, K. E. Huggins, M. Keser, and A. Armstutz, *Science*, **276**, 384, (1997).

[35]. E. W. Jones and J. Gormally, *"Aggregation Processes in Solution"*, Elsevier Sci. Pub. Co., New York, NY, (1983).

[36]. S. Priyanto, G. A. Mansoori and A. Suwono, "Structure & Properties of Micelles and Micelle Coacervates of Asphaltene Macromolecule" in *Nanotechnology Proceed. 2001 AIChE Ann. Meet.*, AIChE, New York, NY, *(2001)*.

[37]. S. Priyanto, G. A. Mansoori and A. Suwono, *Chem. Eng. Science*, **56, 6933**, (2001).

[38]. B. K. Green, L. Schleicher. *U.S. Patent* 2 800 457, (1957).

[39]. J. H. Pacheco-Sanchez and G. A. Mansoori, Paper # 12, Proceed. 2nd Int'l Symp. Colloid Chem. in Oil Prod. (ISCOP '97), 6p, Rio de Janeiro, Brazil, Aug. 31- Sept. 3, (1997).

[40]. J. H. Pacheco-Sanchez and G. A. Mansoori, *Petrol. Sci. & Tech.*, **16**, 3&4, 377, (1998).

[41]. M. Ji, X. Chen, C. M. Wai, and J. M. Fulton, *J. Am. Chem. Soc.* **121**, 2631, (1999).

[42]. A. Cabañas and M. Poliakoff, *J. Materials Chem.*, **11**, 5, 1408, (2001).

[43]. K. M. K. Yu, A. M. Steele, J. Zhu, *J. Materials Chem.*, **13**, 1, 130, (2002).

[44]. G. A. Mansoori, K. Schulz and E. Martinelli, *Bio/Technology*, **6**, 393, (1988).

[45]. T. Y. Kwak and G. A. Mansoori, *Chem. Eng. Sci.*, **41**, 5, 1303, (1986).

[46]. S. J. Park, T. Y. Kwak and G. A. Mansoori, *Int'l J. Thermophysics*, **8**, 449, (1987).

[47]. R. Hartono, G. A. Mansoori and A. Suwono, *Chem. Eng. Sci.*, **56**, 6949, (2001).

[48]. K. P. Johnston, K. L. Harrison, M. J. Clarke, S. M. Howdie, M. P. Heitz, F. V. Bright, C. Carlier, T. W. and Randolph, *Science* **271**, 624, (1996).

[49]. S. Himran, A. Suwono, and G. A. Mansoori, *Energy Sources J.*, **16**, 117, (1994).

[50]. G. A. Mansoori, H. L. Barnes and G. M. Webster, *"Fuels and Lubricants Handbook"*, ASTM Int'l., West Conshohocken, PA, Chap. 19, 525, (2003).

Chapter 8

Positional Assembly of Atoms and Molecules

"Suppose that we wanted to copy a machine, such as a brain, that contained a trillion components. Today we could not do such a thing (even were we equipped with the necessary knowledge) if we had to build each component separately. However, if we had a million construction machines that could each build a thousand parts per second, our task would take only minutes. In the decades to come, new fabrication machines will make this possible. Most present-day manufacturing is based on shaping bulk materials. In contrast, the field called 'nanotechnology' aims to build materials and machinery by placing each atom and molecule precisely where we want it." **Marvin L. Minsky**

Introduction

The controlled and directed organization of molecular building blocks (MBBs) and their subsequent assembly into nanostructures is one fundamental theme of bottom-up nanotechnology. Such an organization can be in the form of association, aggregation, arrangement, or synthesis of MBBs through van der Waals forces, hydrogen bonding, attractive intermolecular polar interactions, electrostatic interactions, hydrophobic effects, etc. The ultimate goal of assemblies of nanoscale molecular building blocks is to create nanostructures with improved properties and functionalities heretofore unavailable to conventional materials and devices. Fabrication of nanostructures demands appropriate methods and molecular building blocks (MBBs).

Through the controlled and directed assembly of nanoscale molecular building blocks one should be able to alter and engineer materials with desired properties. For example, ceramics and metals produced through controlled consolidation of their MBBs have been shown to possess properties substantially improved and different from materials with coarse microstructures. Such different and improved properties include greater hardness and higher yield strength in the case of metals and better ductility in the case of ceramic materials [1,2].

In order to construct specific nanostructures suitable molecular building blocks (MBBs) are required. So far, a number of MBBs are proposed among which amino acids, carbon nanotubes, diamondoids, DNA, fullerene, and graphite are well-known [3]. A molecule should, at least, have two linking groups in order to be considered as MBB. The presence of three linking groups (like in graphite) would lead to a two-dimensional or a tubular structure formation. Presence of four or more linking groups would lead to a three-dimensional structure. The molecules with five linking groups can form a three-dimensional solid and those with six linking groups can be attached together in a cubic structure [1]. Adamantane and buckyballs (C_{60}) are the best candidates for six-linkage group MBBs. It should be pointed out that functionalization of buckyballs with six functional groups is presently possible [4]. The association of linking groups together is necessary for nanostructure assembling. Assembly techniques of this nature envisioned in nanotechnology are of two kinds: positional (or robotic) assembly and self-assembly. In this chapter the positional assembly is introduced. In the following chapter self-assembly is presented.

Positional (or Robotic) Assembly

It has been a long-standing dream of scientists to manipulate individual atoms and molecules in the same way we manipulate macroscopic-sized objects. This is now achieved through positional (robotic) assembly devices.

In olden days there was the talk of alchemy. Now we have positional assembly in hand. We cannot build atoms through positional assembly, but we can make molecules and various materials made of molecules. According to Ralph C. Merkle [5] if we rearrange the atoms in coal we can make diamond, if we rearrange the atoms in sand (and add a few other trace elements) we can make computer chips, and if we rearrange the atoms in dirt, water and air we can make potatoes. While we have not done so any of these, yet, the prospects of doing them are here with the new interrogating and manipulating devices available in nanotechnology. Positional assembly (also called robotic assembly) refers to several methods that are being developed and can be used to create extremely small devices, machines and systems a millionth of a millimeter in size through manipulation and arrangement of atoms and molecules. In positional assembly, the technician is able to see, control and analyze what is taking place in nanoscale as the construction process is carried out.

In positional or robotic assembly each MBB is positioned in a certain coordinate using a robotic arm like an AFM tip [1,2]. The efficiency of this approach is dependent, to a large extent, on the improvements in the field of atomic force microscopy [6,7]. A familiar example of natural positional assembly is protein synthesis control by the ribosome [2].

The major difficulty in manual-design of positional assembly is overcoming the thermal noise, which can cause positional uncertainty [1]. Such thermal noise could be due to the fact that MBBs may not be stiff enough or the devices used for positional assembly may have vibrations due to thermal stress and strain. This problem can be solved, to some extent, by using stiff and rigid MBBs (like diamondoids) and also lowering the temperature during assembly. Assemblers with robotic arms also need to be developed in order to gain control over steric three-dimensional orientations of a molecule [8].

There are two strategies to any positional assembly [1]: (i). Additive synthesis in which MBBs are arranged to construct the desired nanostructure. (ii). Subtractive synthesis in which small blocks are removed from a large building block or a primitive structure to form an eventual structure (like sculpture). In contrast to self-assembly, which will be discussed in the next chapter, positional or robotic assembly is an approach in which each MBB is positioned in a certain coordinate using a robotic arm like an AFM tip [1,2]. The success of positional (robotic) assembly is dependent, to a large extent, on the availability and improvements in the field of scanning probe microscopy (the family AFMs and STMs) [2,6,9,10] as well as use of appropriate MBBs.

There are numerous alternatives for having MBBs, made from tens to tens of thousands of atoms, which provide a rich set of possibilities for parts. Preliminary investigation of this vast space of possibilities suggests that building blocks that could have multiple links to other building blocks (at least three, and preferably four or more) will make it easier to positionally assemble strong and stiff three-dimensional structures. As it will be discussed in chapter 11, adamantane and heavier diamondoids with the general chemical formula $C_{4n+6}H_{4n+12}$, are one such group of MBBs. Adamantane is a tetrahedrally symmetric stiff hydrocarbon that provides obvious sites for four, six or more functional groups.

Assemblers, or positional devices, are applied to hold and manipulate MBBs in positional (robotic) assembly. In fact an assembler would hold the MBBs near to their desired positions using a restoring force, F, which is calculated like a spring force from the following equation [11]:

$$F = s \cdot x,$$

where x stands for distance between a MBB and the desired location (displacement) and s is the spring constant of the restoring force (or stiffness). The relationship between positional uncertainty (mean error in position), $è$, temperature, T, and stiffness, s, is [11]:

$$è^2 = k_B T / s.$$

The product ($k_B T$) denotes thermal noise where k_B is the Boltzmann constant. A scanning probe microscope (SPM) can provide a s value of $10 (\phi [nm])^{-1}$ and thus an $è$ value of $0.02 \phi [nm]$ would be obtained at the room temperature which is a small enough mean error in positions for nanostructural assembly [2].

If the $è$ value becomes very large the positional assembly will not work and one has to resort to using more stiff MBBs or other assembly techniques when possible.

K. Eric Drexler suggested that a universal assembler must be designed which would be able to build almost any desired nanostructure

[10]. However, this may seem impossible at the first glance unless the simple assemblers are originally built and these simple assemblers would build more complicated assemblers and so on, to the point that we can have a universal assembler.

An example of simple assemblers [10] is the "Stewart platform" as shown in Figure 1.

Figure 1. The most basic form of an assembler is the Stewart platform consisting of an octahedron connecting the triangular base to the triangular platform by 3 variable-length struts [10].

In this basic form, the Stewart platform is made of an octahedron shape. The choice of this shape is based on the observation that a polyhedron, all of whose faces are triangular, will be rigid and flexible. One of the two triangular faces is designated as the "platform" and the opposing triangular face is designated as the "base". The six struts of the octahedron that connect the base to the platform are designed so as to be varied in length. Then, changing the lengths of the six struts can vary the orientation and position of the platform with respect to the base. Since there are six struts, and if we are able to independently adjust the length of each strut, the Stewart platform could then be used to position the platform in full six degrees of freedom.

Several molecular assembling devices have been designed and proposed inspiring from the basic Stewart platform shown in Figure 1.

The various methods of improving the range of motion of the platform have been introduced.

K. Eric Drexler proposed an advanced version of Stewart platform [10] as shown in Figure 2. In this design the six adjustable-length edges are either in pure compression or pure tension and are never subjected to any bending force.

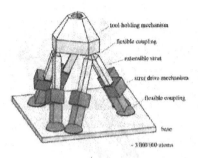

Figure 2. An advanced form of Stewart platform proposed by K. Eric Drexler [10]. The six adjustable-length edges are either in pure compression or pure tension and are never subjected to any bending force, this positional device is quite stiff.

Ralph C. Merkel [11] has proposed a family of positional devices, which provide high stiffness and strength, as do the Stewart platform, but with an increased range of motion and shortening and lengthening of variable struts (see Figure 3).

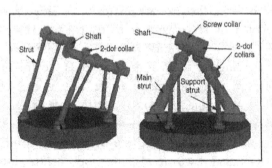

Figure 3. Merkle's positional control devices [11]. Left) Crank - A variation of the Stewart platform in which the platform is replaced with a bent shaft. Right) Double-tripod – A stiff and stable device. The main strut connects to the shaft by a screw collar, which can move along the shaft.

The Stewart platform and its modifications can provide quite higher stiffness than a robotic arm of the present day scanning probe microscopes. However, increased stiffness is usually associated with a decreased range of motion. The double tripod design (Figure 3), provides greater stiffness for a given structural mass than a robotic arm, but has a significantly greater range of motion than a Stewart platform.

Scanning Probe Microscopy

Since scanning probe microscopes are the main components by which positional assembly is presently achieved, we present some basic principles behind these microscopes here.

The idea behind the development of the family of scanning probe microscopes started with the original invention of the Topografiner between 1965 and 1971 by Russell D. Young [13-16] of the United States National Bureau of Standards, now known as the National Institute of Standards and Technology (NIST). The same principle was later improved and used in the scanning tunneling microscope [17] by G. Binnig and H. Rohrer of the IBM Research Laboratory in Zurich, Switzerland. Also in 1986 the invention of the atomic force microscope (AFM) was announced by G. Binnig, et al [18]. In what follows we present the principles behind Topografiner, STM and AFM.

1. Topografiner

It is well known that the electron clouds around the nucleus of metal atoms at a surface of metal extend to a very small distance outside the surface. As a result of that, when a quite sharp needle, which is specially made with a single atom sharpness at its tip, is moved extremely close to a metal surface (a few fractions of nanometer), a strong interaction between the electron cloud on the metal surface and that of the tip atom will occur.

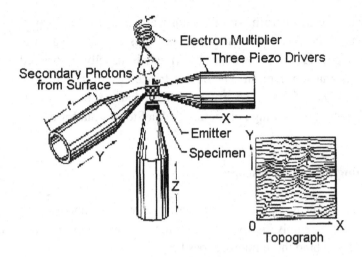

Figure 4. Principle of Topografiner as designed by Russell D. Young [13-16]. This is the basis of all the scanning probe microscopes which were invented later.

Provided a voltage is applied between the metal surface and the needle a quantum mechanical tunneling effect (as presented below) will occur. At a separation of a few atomic diameters between the metal surface and the needle, the tunneling current rapidly increases as the distance between the tip and the surface decreases. This rapid change of tunneling current with distance is a strong function of distance. As a result, when the surface is scanned by the needle tip it can then produce an atomic resolution image of what is on the surface (see Figure 4).

2. Quantum Mechanical Tunneling Effect

Quantum mechanical tunneling effect is due to the fact that electrons have wavelike properties. These waves of electrons do not end abruptly at a barrier but they may decay. Their decay follows an exponential decay function. For the simplest case when we have a rectangular barrier the following expression holds for the electron wave function at a distance S inside the barrier:

$$\psi(s) = \psi(0)e^{-\kappa s} ,$$

and

$$k = \frac{\sqrt{2m(\varphi - E)}}{\hbar}.$$

In the above equations κ is the inverse decay length, ψ is the electron wave probability function, S is thickness of the barrier, E is the electron energy, φ is the barrier height, $\hbar = h/2\pi$, where h is the Plank's constant, and m is a constant parameter which is a function of the nature of the barrier [19]. Depending on the nature and thickness of the barrier (needle tip) electron waves could become thinner after tunneling through the barrier as it is shown in Figure 5. Because of the probabilistic nature of this phenomenon some of the electrons will indeed tunnel through and will appear on the other side of the barrier (tip). When an electron moves though the barrier in this fashion, it is called quantum mechanical tunneling effect.

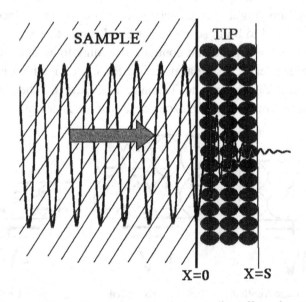

Figure 5. Demonstration of quantum mechanical tunneling effect due to the fact that electrons have wavelike properties. These waves of electrons do not end abruptly at a barrier (tip) but they may decay. Their decay follows an exponential decay function.

The probability of finding an electron at the other side of a barrier of a width S is [19]:

$$|\psi(s)|^2 = |\psi(0)|^2 e^{-2\kappa s}$$

The principle of quantum mechanical tunneling effect, as presented above, is utilized in the design of scanning probe microscopes.

3. Piezoelectric Phenomena

The word "piezo" is a Greek language word, which means "push". Pierre and Jacques Curie brothers originally discovered the phenomena known as piezoelectricity when they were 21 and 24 years old, respectively, in 1880. This effect is creation of electric charge in the sides of certain crystalline materials by squeezing them. Quartz and barium titanate crystals are examples of piezoelectric materials. The effect can be reversed as well (causing compression and expansion of a piezoelectric material) by applying a voltage across its two opposite surfaces. These materials are used to build scanners for STMs and other scanning probe microscopes. They expand and contract according to an applied voltage. A widely piezoelectric material used in SPMs is PZT which is an alloy made of Lead, Zirconium and Titanate. In Figure 6 schematic of the principle behind operation of piezoelectric scanner is demonstrated.

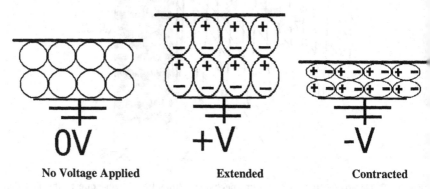

No Voltage Applied   Extended   Contracted

Figure 6. Principle of piezoelectric scanner operation. Electric AC signals applied to conductive areas of the tube will create piezo movement along the three major axes.

4. Scanning Tunneling Microscope (STM)

The STM also works by scanning a very sharp metal wire tip over a surface. By bringing the tip very close to the surface (a few fractions of nanometer), and by applying an electrical voltage ($U \leq 4V$) to the tip or sample, a small electric current (0.01nA-50nA) can flow from the sample to the tip or reverse.

The STM is based on several principles, the most important of which are the following:

(i) The quantum mechanical tunneling effect and related advanced electronics that allows the investigator to see clearly the surface.

(ii) Advanced piezoelectric effect, which allows one to precisely scan the tip with nano-level control.

(iii) A feedback loop which monitors the tunneling current and accurately coordinates the current and the positioning of the tip.

Since the tunneling current reduces exponentially on the distance between the tip and sample, STM is quite sensitive to minute changes in the density of states of the sample. Accordingly, by accurate recording of the lateral position of the tip and its vertical movements it is possible to construct a precise image. In the case of metallic surfaces, these images are considered as topographic images. For semiconductor surfaces the interpretation is not so straightforward due to the fact that the density of states is modulated in space and can change from atom to atom of the semiconductor material. Nevertheless, it has been possible to study the morphology of the semiconductor surfaces during their growth processes and determine the density, size and distribution of islands and also, in some cases, the distribution of atomic species on surface of alloys. The basic parts and comparison of the operations of a scanning tunneling microscope (STM) and an atomic force microscope (AFM) are shown in Figure 7. In 1986, Binnig and Rohrer received the Nobel Prize in Physics for their discovery of atomic resolution in scanning tunneling microscopy.

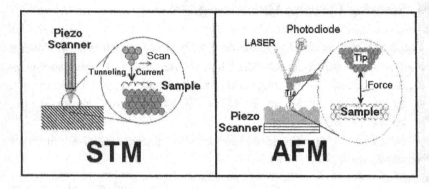

Figure 7. Comparison of the operations of a scanning tunneling microscope (STM) and an atomic force microscope (AFM).

5. Electronics Feedback Loop

An electronic feedback loop is an advanced control system, which is designed so that it constantly monitors the tunneling current and makes adjustments to the tip to maintain a constant tunneling current. The adjustments are automatically recorded by a computer, and through a signal to image processing software, it is illustrated as an image on the computer monitor. Now a days STMs are designed with remarkable capabilities to produce high resolution images on the atomic scale.

6. Atomic Force Microscope (AFM)

Binnig, Quate, and Gerber [18] reported the invention of atomic force microscope (AFM) in 1986. The purpose of their invention was to be able to acquire sample images in the atomic level of, both, conductive and non-conductive samples with a good lateral and vertical resolution. AFM is widely used for observing and analyzing sample surfaces in the atomic scale. The atomic force microscope helps determine possible problems that are emerging in the new technology areas, and is also used to modify and pattern various types of materials on surfaces, for positional nano assembly and other nano applications [20].

With STM it is not possible to scan non-conducting surfaces due to the tunneling current required between the tip and sample during the scan. However, This can be achieved with AFM. Also the AFM's tip makes a soft physical contact with the sample's surface, which will not damage the surface [18]. As it is mentioned above the basic operation of an AFM as compared with a STM is shown in Figure 7.

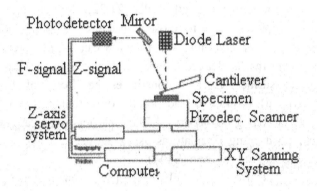

Figure 8. The basic components of an atomic force microscope (AFM).

In Figure 8 the components of an AFM are depicted. According to this figure the laser beam comes from the laser diode and is focused onto the backside of the cantilever. The laser beam reflects from the back of the cantilever and goes to a mirror where the beam reflects and is directed to a four-point photodetector. The sensitive photodetector detects the deflection coming from the laser beam when the cantilever moves as the sample is scanned. Any detection by the photodetector results in a signal that is sent to the feedback loop. The feedback loop will register to move the piezo in the z direction taking the laser beam back to the original position on the photodetector. The sample is scanned with a constant force due to the z piezo motion producing a topographical map of the region scanned.

Applications of STM for Positional Assembly of Molecules

An interesting example of the application of STM for positional assembly is the work of four scientists at the *IBM Almaden Research Center* who discovered a new method for confining electrons to artificial structures at the nanometer length-scale [21,22]. Surface state electrons on Cu(111) were confined to closed structures (corrals) defined by barriers built from Fe adatoms. The notation Cu(111) is referred to the crystal of copper that is cleaved along one of its planes making it having a very smooth surface. Cu(111) constitutes the "floor" of this picture. The word Adatoms is short for "adsorbed atoms" that just stick to the Cu crystal, without being incorporated. The barriers were assembled by individually positioning Fe adatoms using the tip of a low temperature scanning tunneling microscope (STM).

A circular corral of radius 7.13 ϕ [nm] was constructed in this way out of 48 Fe adatoms. This STM image (see Figures 9 and 10) shows the direct observation of standing-wave patterns in the local density of states of the Cu(111) surface. These spatial oscillations are quantum-mechanical interference patterns caused by scattering of the two-dimensional electron gas off the Fe adatoms and point defects.

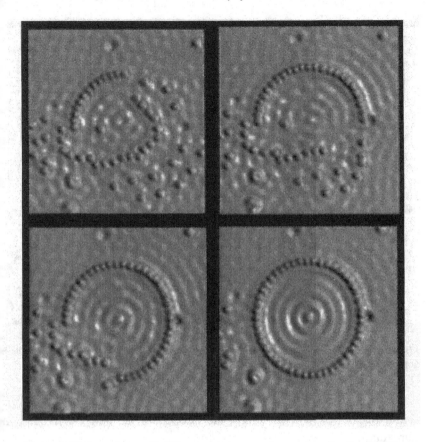

Figure 9. Various stages of building a quantum corral, one atom at a time [9,10].

Another interesting example of application of STM for positional assembly is reported in Figure 11 which shows fullerenes nested within nanotubes, like so many peas in a pod [23]. These nanoscopic peapods which are created by filling the cores of single-wall carbon nanotubes are shown to have tunable electronic properties and they could have far-reaching implications for the fabrication of single-molecule-based devices, such as diodes, transistors and memory elements [23]. Measurements have shown that encapsulation of molecules can

Figure 10. Final STM probe images of Fe adatoms on Cu(111) produced by the scientists at the IBM Almaden Research Center. For details of this work see Refs. [21,22].

dramatically modify the electronic properties of single-wall carbon nanotubes.

It is also shown that such an ordered array of encapsulated molecules can be used to engineer electron motion inside nanotubes in a predictable way. To explore the properties of these novel nanostructures, Hornbacker et al [23] with a low-temperature and high resolution scanning tunneling microscope were able to image the physical structure of individual peapods and map the motion of electrons inside them. They demonstrated that the encapsulated C_{60} molecules modify the local electronic structure of the nanotube without affecting its atomic structure.

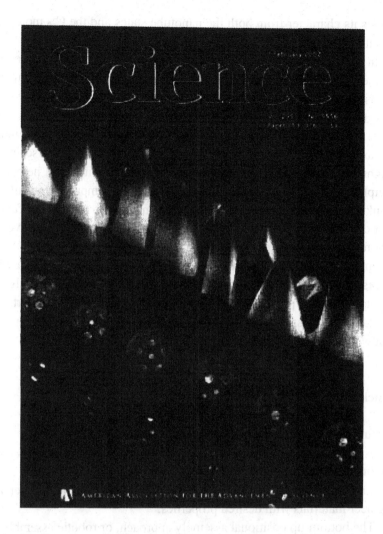

Figure 11. Cover page of February 1, 2002 Science Magazine depicting the atomic structure of a single-wall carbon nanotube "peapod" with superimposed electron waves mapped using a low-temperature scanning tunneling microscope [23].

Measurements and calculations also demonstrate that a periodic array of C_{60} molecules gives rise to a hybrid electronic band, which derives its character from both the nanotube states and the C_{60} molecular orbitals.

In general positional assembly approaches have not been yet advanced enough because of so many limitations in designing suitable MBBs, robotic arms and other associated technical problems. However, with the expected fast pace of activities in nanotechnology in the near future breakthroughs are expected in positional (robotic) assembly. Advancement in this line of nanoscale research could yield scientific insights into the behavior of very small structures, which will help such disciplines as computer chip design. Designers of computer chips like to shrink the features on integrated circuits. In the not too distant future, such developments, by using SPM technology, may eventually pave the way for circuits made from atomic and molecular components. Such extremely small circuits will then be hundreds of times smaller than what is presently used. Such a miniaturization will allow computer designers to put even more processing power onto their chips. That will certainly result in faster, smaller, and lower-electric-power-requiring computers than what is presently available.

Conclusions and Discussion

The ultimate goal of positional assembly is to be able to do the same things with molecules as we can do with macroscopic building blocks like bricks. If we could have sufficient control over the positioning of the right molecular parts in right places then we will be able to alter and engineer materials with desired properties.

The bottom-up positional assembly approach, or robotic assembly, of molecular building blocks, in nanotechnology appears to be an area of exponential growth. While new and improved positional assembly devices are being designed, new positional assembly achievements are continuously announced.

The idea of manipulating and positioning individual molecular building blocks is still new and in most of the achievments, so far, are based on trial and error principles. As the needed tools and devices for positional assembly are improved the technology for positional assembly will mature. As Feynman mentioned in his 1959 classic talk: *"The principles of physics, as far as I can see, do not speak against the possibility of maneuvering things atom by atom."*

Bibliography

[1]. R. C. Merkle, *Nanotechnology*, **11**, 89, (2000).

[2]. R. C. Merkle, *TIBTECH*, **17**, 271, (1999).

[3]. G.A. Mansoori: *United Nations Tech Monitor*, 53, Sep-Oct, (2002).

[4]. K. Hutchison, J. Gao, G. Schick, Y. Rubin and F. Wudl, *J. Am. Chem. Soc.*, **121**, 5611, (1999).

[5]. R. C. Merkle, [http://www.zyvex.com/nano/].

[6]. C. N. R. Rao, A. K. Cheetham, *J. Mater. Chem.*, **11**, 2887 (2001).

[7]. C. R. Lowe, *Current Opinion in Structural Biology*, **10**, 428, (2000).

[8]. K. E. Drexler, [http://www.foresight.org/Updates/Update28/Update28.1.html].

[9]. Christopher R. Lowe, *Current Opinion in Structural Biology*, **10**, 428 (2000).

[10]. K. E. Drexler, *"Nanosystems: Molecular Machinery, Manufacturing, and Computation"* Wiley Interscience, New York, NY, (1992).

[11]. R. C. Merkle, *Nanotechnology*, **8**, 2, 47, (1997). Also at: [http://www.zyvex.com/nanotech/6dof.html]

[12]. K. E. Drexler, [http://www.imm.org/Parts/Parts2.html].

[13]. R. D. Young, *Rev. Sci. Instrum.*, **37**, 275, (1966).

[14]. R. D. Young, *Physics Today*, **24**, 42, Nov., (1971).

[15]. R. D. Young, J. Ward, and F. Scire, *Phys. Rev. Lett.* **27**, 922, (1971).

[16]. R. D. Young, J. Ward, and F. Scire, *Rev. Sci. Instrum.* **43**, 999, (1972).

[17]. G. Binnig and H. Rohrer, *Sci. America*, **253**, 50, (1985).

[18]. G. Binnig, C.F. Quate and C.H. Gerber, *Phys. Rev. Lett.*, **56**, 9, 930, (1986).

[19]. G. C. Schatz and M. A. Ratner, *"Quantum Mechanics in Chemistry"*, Dover Pub. Co., Mineola, NY, (2002).

[20]. M. Wendel, B. Irmer, J. Cortez, R. Kaiser, H. Lorenz, J.P. Kotthaus and A. Lorke, *Supperlattices and Microstructures* **20**, 349, (1996).

[21]. M. F. Crommie, C. P. Lutz, and D. M. Eigler, *Nature*, **363**, 524, (1993).

[22]. M. F. Crommie, C. P. Lutz, and D. M. Eigler, *Science*, **262**, 218, (1993).

[23]. D. J. Hornbacker, S. J. Kahng, S. Misra, B. W. Smith, A. T. Johnson, E. J. Mele, D. E. Luzzi, and A. Yazdani, *Science*, **295**, 828, (2002).

Chapter 9

Molecular Self-Assembly

"As the wind of time blows into the sails of space, the unfolding of the universe nurtures the evolution of matter under the pressure of information. From divided to condensed and on to organized, living, and thinking matter, the path is toward an increase in complexity through self-organization.

Thus emerges the prime question set to science, in particular to chemistry, the science of the structure and transformation of matter: how does matter become complex? What are the steps and the processes that lead from the elementary particle to the thinking organism, the (present!) entity of highest complexity?

And there are two linked questions: an ontogenetic one, how has this happened, how has matter become complex in the history of the universe leading up to the evolution of the biological world, and an epigenetic one, what other and what higher forms of complex matter can there be to evolve, are there to be created?

Chemistry provides means to interrogate the past, explore the present, and build bridges to the future." **Jean-Marie Lehn**

Introduction

Self-assembly is a process in which a set of components or constituents spontaneously forms an ordered aggregate through their global energy minimization. A vast number of macromolecules (like proteins, nucleic acid sequences, micelles, liposomes, and colloids) in nature adapt their final folding and conformation by self-assembly processes. There exist many examples in the literature of natural self-assemblies which occur spontaneously due to the forces of nature. Examples of such natural self-assemblies are observed at all levels, from the molecular to macromolecular scale and in various living systems. Protein folding, diamondoids as host-gust chemistry molecular receptors, DNA double helix, formation of lipid bilayers, formation of micelles, micelle coacervates and steric colloids are some examples in which self-assembly is the dominant phenomenon [1-7].

Nanotechnology self-assembly encompasses a wide range of concepts and structural complexities, from the growth of crystals to the reproduction of complete biological entities. It is the aim of nanotechnology to master and design well controlled self-assemblies starting from various MBBs. By utilizing the nature's help in such self-assemblies one could form and produce various nano structures and then larger systems and materials with the desired physicochemical properties. Such larger heterogeneous aggregates should be able to perform intricate functions, or constitute new material forms, with unprecedented properties. This could lead to the creation of objects able to find their place on particular substrates or within given infrastructures.

Directed self-assembly to design the desired synthetic nanostructures from MBBs is the main goal of nanotechnology. Undoubtedly to achieve this goal it is necessary to utilize the knowledge about intermolecular interactions between MBBs, nanostructures steric arrangements, computer based molecular simulations, and biomimetic [7-9]. Biomimetic is defined as making artificial products that mimic the natural ones based on the study of structure and function of biological substances.

The Five Factors Responsible for Self-Assembly

According to Whitesides and Boncheva [1] five factors are responsible for the success of a self-assembly process. They include: (1) Molecular building blocks; (2) Intermolecular interactions; (3) Reversibility (or adjustability); (4) Molecular mobility (due to mixing and diffusion); (5) Process medium (or environment).

(1). The Role of Molecular Building Blocks (MBBs) in Self-Assembly: The self-assemblies of major interest in nanotechnology are mostly among larger size molecules, known as the molecular building blocks (MBBs), in the size ranges of $1-100\phi$ [nm]. This is because larger well-defined MBBs offer greater level of control over other molecules, and over the interactions among them that makes fundamental investigations especially tractable [1]. The molecular building blocks and their desired properties are discussed in Chapter Eleven with a detailed discussion about diamondoids as one of the most versatile and promising category of molecular building blocks.

(2). The Role of Intermolecular Interactions in Self-Assembly: Forces responsible for self-assembly are, generally, weak non-covalent intermolecular forces including hydrogen bonding, electrostatic, van der waals, polar, hydrophobic and hydrophilic interactions. In Chapter 2 a detailed presentation of interatomic and intermolecular forces and potentials important in nanotechnology is given.

To assure consistency and stability of the whole self-assembly complex the number of such weak interactions for self-assembly formation in each region of molecular conformation must be rather high. An example of a stable self-assembly made up of weak interactions is the second helical beta-sheet structure of proteins formed through hydrogen bonding [1,10] as shown in Figure 1.

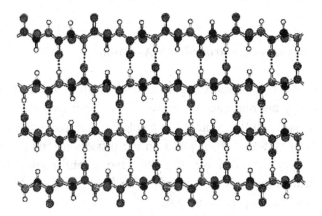

Figure 1. Many hydrogen bonds (dotted lines) would form if two parts of a peptide chain would pair up parallel with one another as shown in this figure. This formation is called a beta-structure, and it often referred to as a beta sheet. Silk fibroin, the protein that makes up silk is an example of beta-sheet.

(3). Reversibility: The ongoing and other envisioned self-assembly processes in nanotechnology are controlled, but spontaneous, processes during which the MBBs associate and produce the desired ordered assemblies or aggregates [5-7]. To achieve spontaneity it is necessary to run the process reversibly.

(4). Molecular Mobility: Due to its dynamic nature, a self-assembly process needs a fluid medium to proceed. Liquids, gases, supercritical fluids and the fluid side of the interfaces between a solid and a fluid phase are the possible environments for self-assembly. In all such cases, to achieve self-assembly, dynamic exchanges toward reaching the minimal energy level has to take place [1].

(5). Process Medium: A self-assembly is largely influenced by the environment in which it occurs. A molecular aggregate which is formed by the self-assembly process, is generally an ordered array which is thermodynamically the most stable conformation. Self-assembly occurs in liquid medium, in gaseous medium (including dense gases - supercritical fluid medium), near the interface between a solid and a fluid, or at the interface between a gas and a liquid [3,4,11-13].

Some Examples of Controlled Self-Assemblies

The principles behind achieving successful controlled self-assemblies in the laboratory are presented above. Now it is appropriate to introduce a number of successfully achieved examples of self-assemblies. There are two kinds of self-assembles based on the process medium. (A). Those which occur on the interface of a fluid and a solid phase. (B). Those which occur in the bulk fluid phase. The fluid phase can be a liquid, a vapor or a dense (supercritical) gas.

(A). Self-Assembly Using Solid Surfaces – Immobilization Techniques

There are a number of self-assembly methods which are achieved in the laboratory using fluid medium as their association environment and a solid surface as their assembling nucleation and growth. An important first step in such self-assemblies is the immobilization of the assembly seed on solid surfaces. There are a number of immobilization techniques used for this purpose, some of which are reported here.

Immobilization of molecules as assembly seed on solid supports, which need to be used for self-assembly, can be achieved via covalent or non-covalent bonds between the molecule and the surface. Covalent bonds produce irreversible, thus stable, immobilization at all stages. Immobilization through non-covalent bonds is a reversible process and unstable at the onset of the process. However, it achieves stability upon appreciable growth of the assembly process.

A more common covalent bond used for immobilization is a sulfide bond and a noble metal. One such example is the covalent bond between thiol-bearing molecules (like alkane-thiol chains or proteins with cystine in their structure) and gold [13]. Some common non-covalent bonds used for immobilization are the following three kinds of bond [14]: (i). Affinity coupling via antibodies. (ii). Affinity coupling by biotin-streptavidin (Bio-STV) system and it's modification. (iii). Immobilized metal ion complexation (IMIC)

In the following sections these three kinds of non-covalent bonds are explained in more detail.

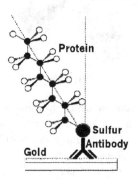

Figure 2. Affinity coupling via antibodies: For direct immobilization, the purified antibodies are attached to gold substrates by using a bifunctional reagent with the thiol group on one side.

(A-1). Affinity Coupling via Antibodies: Antibody is a protein produced by the body's immune system in response to a foreign substance (antigen). Antibodies are glycoproteins that are called immunoglobulins that are found in the blood and tissue fluids produced by cells of the immune system (see Figure 10 in Chapter 1). For direct immobilization, the purified antibodies are attached to gold substrates by using a bifunctional reagent with the thiol group on one side (see Figure 2). Highly oriented antibody immobilization can be achieved by assembling protein on gold surfaces via an introduced cystein-residue [14].

(A-2). Affinity Coupling by Biotin-Streptavidin (Bio-STV) System and Its Modification: Avidin is a basic glycoprotein of known carbohydrate and amino acid content. Avidin combines stoichiometrically with biotin (a vitamin B), which is a small molecule with high affinity to avidin. Streptavidin (STV) is a tetrameric protein, and like avidin, has four high affinity binding sites for biotin (see Figure 3).

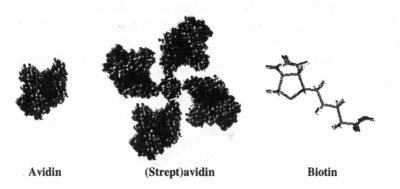

| Avidin | (Strept)avidin | Biotin |

Figure 3. Structures of Avidin (a basic glycoprotein with high affinity to biotin), (Strept)avidin (a protein with four high affinity binding sites for biotin) and Biotin ($C_{10}H_{16}N_2O_3S$). Biotin can be conjugated to a variety of biological molecules, including antibodies, and many biotin molecules can be attached to a single molecule of protein. The biotinylated protein can thus bind to more than one molecule of avidin.

STV is a protein homologous to the avidin of yolk, produced by a strain of bacteria. STV binds very tightly to biotin, producing biotin-streptavidin (Bio-STV) complex. Furthermore streptavidin is not a glycoprotein and therefore does not bind to lectins as does avidin. The physical properties of streptavidin therefore make this protein much more desirable for use in immobilization systems than avidin.

Because of its high affinity constant between the interacting partners, the biotin–avidin system is one of the most prominent systems used for immobilization. Affinity coupling by Bio-STV system is one of the most efficient approaches to construct semi-synthethic nucleic acid-protein conjugates [15]. Bio-STV affinity constant is about 10^{14} dm^{3}mol^{-1}, which is the strongest known ligand-receptor interaction [16]. This method provides the possibility for immobilization of many molecules [17].

(A-3). Immobilized Metal Ion Complexation (IMIC): The IMIC method is based on the non-covalent binding of a biomolecule by complex formation with metal ions. Such metal ions are immobilized by chelators (molecules that bind to metal ions), like iminodiacetic acid [HOOC-CH$_2$NH-CH$_2$-COOH] or nitrilotriacetic acid [HOOC-CH$_2$-N-(CH$_2$-COOH)$_2$] also called NTA (see Figure 4).

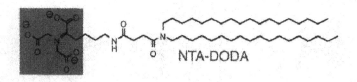

Figure 4. An example of the concept of immobilized metal ion complexation (IMIC), θ's stand for divalent metal ions, NTA stands for nitrilotriacetic acid (chelator), and DODA stands for dimethyl-dioctadecylammonium which is a surfactant sticking to the solid surface (from [14]).

In contrast to other affinity-based immobilization techniques, the interaction is based on low-molecular weight compounds. First, the chelator is attached to a surface, followed by loading with divalent metal cations (e.g. Ni^{2+}, Zn^{2+}, Cu^{2+}). The tetradental ligand NTA forms a hexagonal complex with the central Me^{2+}, occupying four of the six coordination sites. Two sites are then available for binding biomolecules with electron donating groups coordinatively [14,18].

Crystalline bacterial cell surface layers can be exploited for immobilization using appropriate chemical and physical processes [19]. DNA oligomers can be used for site-selective immobilization of macromolecules. This is quite noteworthy, because it becomes feasible to direct self-assembly process on a solid surface via DNA directed immobilization (DDI) [20]. Firstly, a DNA strand tags the desired site of the molecule and then the complementary DNA strand is fixed on a solid surface. Thus, completely specific DNA hybridization is exploited to immobilize macromolecules with controlled steric orientation. This method has been successfully performed to immobilize gold nanoparticles [21].

Figure 5 depicts the role of DNA microarrays for site-selective immobilization of oligonucleotide-modified gold nanoparticles. To achieve this, a "capture oligonucleotides", cA, is first attached to a solid glass support to produce DNA microarrays. On the other side, the gold nanoparticles are tagged by several oligonucleotides, A, which are complementary to the capture oligonucleotide cA.

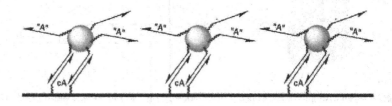

Figure 5. DNA microarrays can be exploited to immobilize oligonucleotide-modified gold nanoparticles site-selectively. A "capture oligonucleotides", cA is first attached to a solid glass support to make DNA microarrays. On the other side, the gold nanoparticles are tagged by several oligonucleotides, A, which are exclusively complementary to the capture oligonucleotide cA. Through the use of a fluorescent probe the successful DDI (DNA directed immobilization) of gold nanoparticles were confirmed [21]. In figure above the circles represent the gold nanoparticle, the solid line represents the solid glass support, the arrowheads represent the complementary DNA ends.

Such DDI techniques allow for highly efficient, reversible and site-selective functionalization of laterally microstructured solid supports

These methods are particularly suitable for fabrication of reusable biochips and other miniaturized sensor devices containing biological macromolecules, enzymes and imunoglobulins [21].

(A-4). Self-Assembled Monolayer (SAM): A self-assembled monolayer (SAM) is defined as a two-dimensional film with the thickness of one molecule that is attached to a solid surface through covalent bonds. There are many applications in nanotechnology for SAM including nanolithography, modification of adhesion and wetting properties of surfaces, development of chemical and biological sensors, insulating layers in microelectronic circuits and fabrication of nanodevices just to name a few. Various examples of protein SAM production methods [13,22] are presented below.

(A-4-1). Physical adsorption: This approach is based on adsorption of proteins on such solid surfaces like carbon electrode, metal oxide or silica oxide. The adsorbed proteins constitute a self-assemble monolayer (SAM) with random orientations. Orientation control may be improved by modification of the protein and surface. This is demonstrated by Figure 6 (a).

(A-4-2). Inclusion in polyelectrolytes or conducting polymers: Polyelectrolytes or conducting polymers could provide a matrix in which the proteins are trapped and attached, or adsorbed, to the surface. This is demonstrated by Figure 6 (b).

(A-4-3). Inclusion in SAM: By using thiolated hydrocarbon chains, it is possible to produce a membrane-like monolayer on a noble metal, through which proteins can be located with non-specific orientations. Utilization of chains with different lengths (depressions and holes) would result in a SAM with definite topography that can give a specific orientation to the proteins. This is demonstrated by Figure 6 (c).

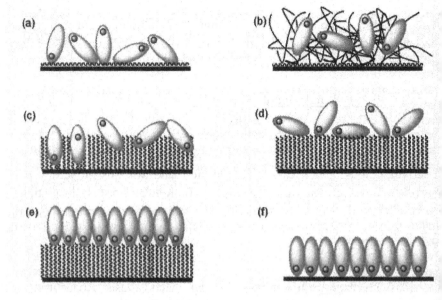

Figure 6. Various methods for production of self-assembled monolayers (SAM) of proteins. (from [13,22]). (a) Physical adsorption; (b) Inclusion in polyelectrolytes or conducting polymers; (c) Inclusion in SAM; (d) Non-oriented attachment to SAM; (e) Oriented attachment to SAM; (f) Direct site-specific attachment to gold.

(A-4-4) Non-oriented attachment to SAM: In this approach the chains which form self-assemble monolayer (SAM) possess a functional group at their ends and react non-specifically with different parts of a protein. So, the orientation is random. This is demonstrated by Figure 6 (d).

(A-4-5) Oriented attachment to SAM: The principles is like the previous approach but the functional group here interacts specifically just with one domain or part of a given protein and hence a definite orientation is obtained. The structure of proteins can be chemically or genetically modified for this purpose. This is demonstrated by Figure 6 (e).

(A-4-6) <u>Direct site-specific attachment to gold:</u> This is achieved by attachment of a unique cystine to gold surface. In this case the orientation can be completely controlled. This is demonstrated by Figure 6 (f).

(A-5). <u>Strain Directed Self-Assembly</u>

Strain directed self-assembly is applicable for fabrication and interconnection of wires and switches. In this approach a lithographically defined surface is impregnated with a strain of a compositionally controlled precipitation agent. It is possible to introduce a functional group to the substrate, which would couple with the surface functionality [23]. This method can be used, for example, for semiconductor construction [24] where it is necessary to immobilize system components on a solid support in order to gain total control over the progress and completion of the self-assembly process.

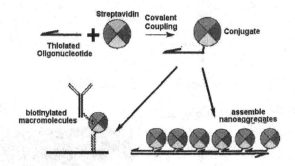

Figure 7. Schematic representation of DNA directed assembly. Employing Bio-STV systems makes it possible to attach a variable and extensive range of macromolecules to the DNA tags in order to achieve DDI and after that DNA directed assembly. A thiolated oligonucleotide is coupled with a streptavidin (STV) molecule to form a conjugate. This can be used either as an auxiliary tool in DDI of different biotinylated macromolecules. or to fabricate and assemble nanoaggregates. In the first case one valence of tetravalent streptavidin is occupied by a biotinylated antibody and the second valence by a thiolated DNA tag in order to immobilize antibodies. In the second case the DNA microarray containing complementary sequences is used as a solid support for DDI and further assembly of tagged nanomodules to form nanoaggregate. This method was first used for site-selective immobilization of proteins and later to assemble nanocrystall molecules from gold nanoclusters. [25].

(A-6). DNA Directed Self-Assembly

DNA can be used, both, for site-selective immobilization and as a linker and thus provides a scaffold for nanostructure assembling. Nucleic acid-protein conjugate synthesis and utilization of specific interactions between two complementary DNA strands, antigen- antibody and Bio-STV, can bring about powerful tools to direct the mode of nanostructure modules attachment. Recent improvements in exerting genetic engineering techniques to the immobilized DNA sequences on a gold surface – like ligation, PCR (Polymerase Chain Reaction), and restriction digestion – provide even more control over self- assembly process [26]. Figure 7 illustrates utilization of DNA- STV conjugates in DNA directed assembly to construct nanostructures. Figure 8 represents usage of biotin-streptavidin system in DNA-directed self- assembly.

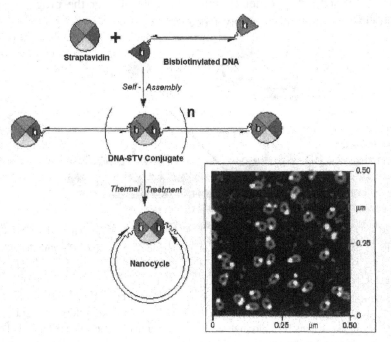

Figure 8. Oligomeric DNA-STV conjugates self-assembles from 5´,5´-bisbiotinylated DNA sequences and streptavidin (STV) as a linker. Thermal treating of DNA-STV conjugate would lead to formation of nanocycles which their AFM image has been shown at the bottom (From [27]).

Such a method can be applied to inorganic nanocrystal molecules. DNA can be also employed for templated synthesis and its example is silver nanowire construction using DNA backbone [28]. DNA has also been utilized to construct one, two or three-dimensional frameworks [25]. Figure 9 depicts such structures.

Figure 9. The "sticky ended" DNA strands associations are used to construct DNA-made nanostructures like "truncated octahedron" (left) and a DNA molecule with the connectivity of a cube (right). The rigidity of the molecules can be improved to some extant by using DNA crossover molecules (From [29]).

It should be noted that DNA oligomer-tagged nanomodules could be attached together with favorite regioselectivity and in a controlled way. Detaching of an assembled nanostructure from its solid support leads to its folding and formation of an eventual conformation, the shape of which depends on the environmental conditions.

(A-7). Self-Assembly on Silicon Surfaces

Employing silicon surfaces for self-assembly is of more importance in the micro/nanoelectronics than other fields. Molecules and metals can be selectively deposited on a silicon template via such a method [30]. This resembles patterning techniques to some extent (silicon based microlithography). Some other substances have been proposed for use in templated self-assembly like S-layer (single layer) proteins [19], diatom frustules [31], etc.

(B). Self-Assembly in Fluid Media

Examples of self-assembly in a liquid, or in a dense-gas (supercritical fluid) phase, medium are the following:

(i) Self-assembly of polymerizable amphiphiles [3,4,32-34], polymer and lipid utilization to form ordered aggregates. Such ordered aggregates can be in the forms of monolayers, bilayers, micelles, micelle-coacervates (Figure 10) or nanoparticles.

Figure 10. An example of micellization of asphaltene macromolecules and self-assembly of the resulting micelles to form micelle-coacervates, all in a liquid medium [3,4].

(ii) Fluidic self-assembly using a template surface and employing liquid stream to select appropriate components, is another approach [1]. Figure 11 illustrates the general procedure in fluidic self-assembly.

(iii) Dynamic combinatorial libraries (DCLs) which have an outstanding position in the combinatorial chemistry, as it will be presented in Chapter 10.

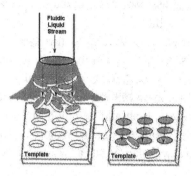

Figure 11. General principle of fluidic self-assembly using a template surface and employing a liquid stream.

(iv) Design of molecular cages through host-guest chemistry as it will be discussed in Chapter 10.

Conclusions and Discussion

In this chapter the factors responsible for, and categories of, self-assembly are presented and various examples of successful self-assemblies are introduced. Based on the material presented above we may conclude that self-assembly is a method of integration in which the components spontaneously assemble, typically by bouncing around in a solution or gas phase until equilibrium is reached (entropy is maximized and free energy is minimized) and a stable structure is generated.

According to Lehn [35,36] self-organization (which is the same as self-assembly) is the basic process (or driving force) that led up to the evolution of the biological world from inanimate matter. Understanding, inducing, and directing self-assembly is key to unraveling the progressive emergence of bottom-up nanotechnology.

By understanding self-assembly we will be able to learn the role of various intermolecular interaction forces which govern a given self-assembly. In order to induce and direct a desired self-assembly, it is also necessary, to be able to model and predict what happens under a certain process condition to an assembly of molecules and what directions they take, if any, to self-assemble.

While self-organization is crucial to the assembly of biomolecular nanotechnology, it is also a promising method for assembling non-biological atoms and molecules, and especially molecular building blocks, towards creating materials with preferred behavior and properties as well as precise devices. It should be emphasized that self-assembly is by no means limited to molecules or the nanoscale. It can be carried out on just about any scale, making it a powerful bottom-up assembly method.

Bibliography

[1]. G. M. Whitesides and M. Boncheva, *PNAS*, **99**, 8, 4769, (2002).

[2]. R. P Sijbesma and E. W. Meijer, *Current Opinion in Colloid & Interface Science*, **4**, 24, (1999).

[3]. S. Priyanto, G. A. Mansoori and A. Suwono, "Structure & Properties of Micelles and Micelle Coacervates of Asphaltene Macromolecule" in *Nanotechnology Proceed. 2001 AIChE Ann. Meet.*, AIChE, New York, NY, (2001).

[4]. S. Priyanto, G. A. Mansoori and A. Suwono, *Chem. Eng. Science*, **56**, 6933, (2001).

[5]. J. M. DeSimone and J. S. Keiper, *Current Opinion in Solid State and Materials Science*, **5**, 4, 333, (2001).

[6]. R. Fiammengo, M. Crego-Calama and D. N. Reinhoudt, *Current Opinion in Chemical Biology*, **5**, 660, (2001).

[7]. H. Ramezani and G. A. Mansoori, *"Diamondoids as Molecular Building Blocks for Nanotechnology* (to be published).

[8]. H. Rafii-Tabar and G. A. Mansoori, *"Interatomic Potential Models for Nanostructures"* in *"Encyclopedia of Nanoscience & Nanotechnology"*, American Sci, Pub., Stevenson Ranch, CA, **4**, pp.231-248, (2004).

[9]. K. Esfarjani and G. A. Mansoori, *"Statistical Mechanical Modeling and its Application to Nanosystems"*, in *"Handbook of Computational Nanoscience and Nanotechnology"*, American Sci, Pub., Stevenson Ranch, CA, (to appear).

[10]. R. C. Merkle, *TIBTECH*, **17**, 271, (1999).

[11]. J. D. Holmes, D. M. Lyons and K. J. Ziegler, *Chemistry* (Weinheim An Der Bergstrasse, Germany), **9**, 10, 2145, (2003).

[12]. Y. Kikuchi, T. Fukuda, S. Shishiguchi, K. Masuda, and N. Kawakami, *Proc. SPIE Int. Soc. Opt. Eng.* 4688, 896, (2002).

[13]. G. Gilardi and A. Fantuzzi, *Trends in Biotechnology*, **19**, 11, (2001).

[14]. K. Busch and R. Tampe, *Reviews in Molecular Biotechnology*, **82**, 3, (2001).

[15]. C. M. Niemeyer, *Reviews in Molecular Biotechnology*, **82**, 47, (2001).

[16]. P. C. Weber, D. H. Ohlendorf, J. J. Wendoloski and F. R. Salemme, *Science*, **243**, 85, (1989).

253

[17]. P. S. Stayton, A. S. Hoffman, N. Murthy, C. Lackey, C. Cheung, P. Tan, L. A. Klumb, A. Chilkoti, F. S. Wilbur and O. W. Press: *J. Controlled Release*, **65**, 203, (2000).

[18]. D. G. Castner, B. D. Ratner, *Surface Science*, **500**, 28, (2002).

[19]. U. B. Sleytr, M. Sara, D. Pum and B. Schuster, *Prog. Surface Sci.*, **68**, 231, (2001).

[20]. S. Peschel, B. Ceyhan, C. M. Niemeyer, S. Gao, L. Chi and U. Simon, *Materials Sci. and Eng.* C, **19**, 47, (2002).

[21]. C. M. Niemeyer, B. Ceyhan, S. Gao, L. Chi, S. Peschel and U. Simon, *Colloid Polym Sci*, **279**, 68, (2001).

[22]. S. Ferretti, S. Paynter, D A. Russell, K. E. Sapsford and D. J. Richardson, *Trends Anal. Chem.*, **19**, 530, (2000).

[23]. R. A. Kiehl, M. Yamaguchi, O. Ueda, N. Horiguchi and N. Yokoyama, *Appl. Phys. Lett.*, **68**, 478, (1996).

[24]. C. N. R. Rao, A. K. Cheetham, *J. Mater. Chem.*, **11**, 2887, (2001).

[25]. C. M Niemeyer, *Current Opinion in Chemical Biology*, **4**, 609, (2000).

[26]. J. H. Kim, J. A. Hong, M. Yoon, M. Y. Yoon, H. S. Jeong and H. J. Hwang, *J. Biotechnology*, **96**, 213, (2002).

[27]. C. M. Niemeyer, M. Adler, S. Gao and L. F. Chi, *Angew Chem Int Ed*, **39**, 3055, (2000).

[28]. E. Braun, Y. Eichen, U. Sivan and G. Ben-Yoseph, *Nature*, **391**, 775, (1998).

[29]. N. C. Seeman, H. Wang, X. Yang, F. Liu, C. Mao, W. Sun, L. Wenzler, Z. Shen, R. Sha, H. Yan, M. H. Wong, P. Sa-Ardyen, B. Liu, H. Qiu, X. Li, J. Qi, S. M. Du, Y. Zhang, J. E. Mueller, T. J. Fu, Y. Wang, and J. Chen, 5th Foresight Conf. on Molecular Nanotechnology:
<http://www.foresight.org/Conferences/MNT05/Papers/Seeman/>, (1997).

[30]. F. J. Himpsel, A. Kirakosian, J. N. Crain, J. L. Lin and D. Y. Petrovykh, *Solid State Commun*, **117**, 149, (2001).

[31]. J. Parkinson and R. Gordon, *TIBTECH*, **17**, May, (1999).

[32]. T. Koga, S. Zhou, and B. Chu, *Appl. Opt.*, **40**, 4170, (2001).

[33]. G. D. Wignall, D. Chillura-Martino and R. Triolo, *Science*, **274**, 5295, 2049, (1996)

[34]. S. A. Miller, J. H. Ding and D. L. Gin, *Current Opinion in Colloid and Interface Sciences*, **4**, 338, (1999).

[35]. J.-M. Lehn, *PNAS*, **99**, 8, 4769, (2002).

[36]. J.-M. Lehn, in *"Supramolecular Science: Where It Is and Where It Is Going"*, eds. R. Ungaro and E. Dalcanale, Kluwer, Dordrecht, The Netherlands, pp. 287-304, (1999).

Chapter 10

Dynamic Combinatorial Chemistry

"Everything we know and do. Was at one time wild and new!" **Donald J. Cram**

Introduction

Combinatorial chemistry, in general, means a route to discover promising compounds and self-assemblies that is better, cheaper and faster than other existing methods. Dynamics combinatorial chemistry (DCC) is considered to be an evolutionary approach for bottom-up nanotechnology. For development of a DCC scheme it is necessary to put together a dynamic combinatorial library (DCL) of intermediate components, which upon addition of templates, will produce the expected molecular assembly. In the formation of DCC the concept of molecular recognition becomes important. A complementary approach to DCC is the science of host-guest chemistry. In this chapter we introduce the principles of DCC along with DCL, molecular recognition and host-guest chemistry.

The science of "combinatorial chemistry" refers to using a combinatorial process of experiments to generate sets of molecular assemblies (compounds) from connections between sets of building blocks. It also refers to the high-throughput synthesis and screening of chemical substances to produce and identify new agents with useful properties.

A combinatorial chemistry includes, first, the synthesis of chemical compounds as ensembles (libraries) and, then, the screening of those libraries to search for compounds with certain desirable properties.

Presently combinatorial chemistry is exploited as a basic research tool to determine the function of enzymes and to identify new enzyme inhibitors. Combinatorial chemistry is considered as a potentially speedy route to new self-assemblies in nanotechnology as well as discovery of new drugs, supramolecular assemblies and catalysts [1-3].

There exist two kinds of combinatorial chemistry. They are the traditional combinatorial chemistry and dynamic combinatorial chemistry. The major distinction between the dynamic and traditional combinatorial chemistries is that in the dynamic one the weak, but reversible, non-covalent bonds link the molecular building blocks together while in the traditional one the interactions are basically a result of strong, and irreversible, covalent bonds. The natures and differences between the covalent and non-covalent bonds are discussed in Chapter 2.

In the traditional combinatorial chemistry, a static mixture of fixed aggregates is formed and a template (ligand) selects its best binder without amplification of that binder. In a dynamic combinatorial chemistry a dynamic mixture is obtained which after addition of the template its composition and distribution changes and the best binder to the template would be the only predominant product.

A template (or ligand) in combinatorial chemistry is referred to a molecule, an ion or a macromolecule which interacts with the components of the combinatorial chemistry and it alters the distribution of the products of an ongoing reacting system towards a desired aggregate, macromolecule or product. An example of a template is a DNA that serves as a pattern for the synthesis of a macromolecule, such as RNA. In Figure 1 a schematic comparison between dynamic and traditional combinatorial chemistries is demonstrated.

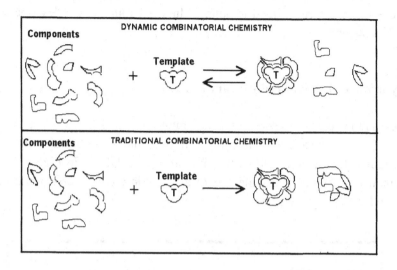

Figure 1. A schematic comparison between dynamic and traditional combinatorial chemistries. In a DCL after addition of a template a dynamic mixture is obtained which is reversible and in equilibrium with the components. In contrast, in a traditional library a static mixture forms which can not be reversed. Hence, after addition of template no change in aggregate concentration distributions occures.

The goal of self-assembly through the "high-throughput" dynamic combinatorial chemistry is to screen and synthesis various groups and classes of chemical substances to produce new materials with desired properties. For such synthesis we must use weak and easily reversed non-covalent interatomic and intermolecular interaction bonds. Non-covalent interactions include hydrogen bonds, van der Waals interactions, electrostatic interactions, hydrophobic effects, polar interactions, etc.

Self-assembly through dynamic combinatorial chemistry (DCC) is a rapidly emerging field, which offers a possible alternative to the other approaches of molecular assembly. Many interesting improvements in the field of self-assembly through dynamic combinatorial chemistry have been reached during the recent years.

In dynamic combinatorial chemistry (DCC) the building blocks, with properties suitable for forming a self-assembled entity, are selected from a dynamic combinatorial library (DCL) and are assembled in the presence of a target called template. In what follows, we describe the DCL concept.

Dynamic Combinatorial Library (DCL)

DCL is the central concept in self-assembly through DCC. DCL refers to a set of intermediate compounds, in dynamic equilibrium with the building blocks, produced as the first stage of a DCC. In Figure 2 we demonstrate symbolically the role of a dynamic combinatorial library in a combinatorial chemistry.

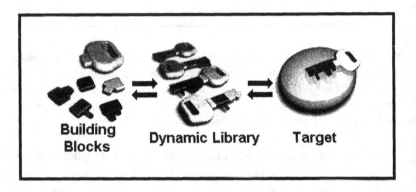

Figure 2. Symbolic demonstration of Dynamic Combinatorial Library (Courtesy of Atlantus Pharmaceuticals).

A combinatorial library may consist of a collection of pools or sub-libraries. The "Chemset notation" is generally used to describe the composition of a DCL. Chemset is a collection of two or more library members, building blocks, or reagents.

The components in a DCL interact through weak non-covalent bonds. As a result creation of all possible reversible assemblies composed of these components is, in principle, achievable. The dynamic nature of a DCL is because the interactions between the components are all reversible and in equilibrium and, in fact thermodynamics, and not kinetics, controls the distribution of a DCL of different assemblies.

Thus, a DCL can be responsive to various external forces, which are exerted on it. Specifically, distribution of aggregates in a DCL could change with variation of its thermodynamic conditions and depending on the nature of the template added to the system (see Figure 3).

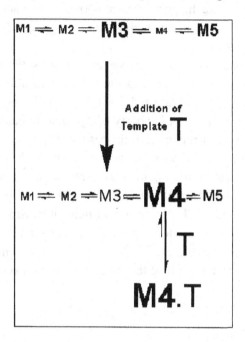

Figure 3. A chart, which depicts DCL redistribution after addition of a template (**T**). The letters size represents the concentration of each member (**M**) in the library (modified from [2]).

In equilibrium state, and before addition of a template, there are many possibilities for components of a DCL to interact with each other and form diverse aggregates via weak non-covalent interactions. After addition of a template to the DCL system the re-distribution would occur. As a result only the concentration of those aggregates or assemblies, which would best fit to the template, will increase and become stabilized due to presence of template.

The amplification of a certain product could result only due to reversible shift of other reactions towards formation of that product provided equilibrium conditions (minimization of energy and maximization of entropy) would dictate that. Hence, the system approaches towards the most stable binding assembly to the template and the concentration of unstable aggregates would diminish. In Figure 3 the role of addition of template which has affinity to one of the aggregates (**M4**) is demonstrated. In principle, components interact spontaneously with each other to produce a wide range of aggregates with different shapes and properties.

After a template addition to the system, the most thermodynamically stable aggregate in interaction with the template would be amplified as the best binders to the template and the other possible aggregates become the minor products. This DCL ability of dynamic redistribution denotes existence of molecular recognition in these systems. By using different suitable templates, it is possible to obtain very different aggregate concentration distributions. Figure 4 presents a real example of a DCL system before and after template (lithium ion, Li^+) addition [2].

In short a proper DCL design would make it possible the pairing of the most appropriate binder to a template (or ligand to receptor).

The reversible reactions, which are generally considered in the design of DCL systems, include the following reaction systems [1]:

disulfide exchange,
ester exchange,
hydrazine exchange,
hydrogen bond exchange,
metal-ligand coordinations,
olefin metathesis,
oximes exchange,
peptide bond exchange,
transamination,
transesterfication and
transimination.

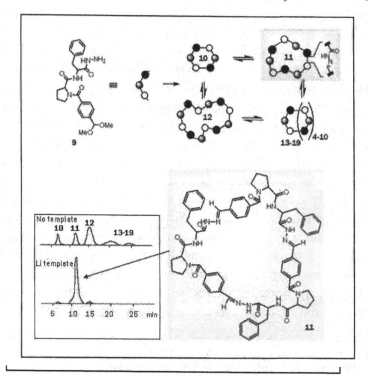

Figure 4. An example of DCL which consists of hydrazone-based pseudopeptide macrocycles [2]. The HPLC chromatogram indicates that there are so many compositions possible before addition of template, namely Li^+, (upper trace) but after addition of Li^+ template the most thermodynamically stable composition becomes predominant (lower trace).

In the above list only disulfide exchange and olefin metathesis can operate under physiological conditions [1,3]. In the case of peptide bond exchange reactions a library of different peptides is obtained using low specificity proteases under suitable conditions. In appropriate conditions, the utilized enzymes are able to catalyze both hydrolysis and synthesis reactions [4].

Separation of the major aggregate formed due to template is one of the difficulties in the design of DCL systems. Affinity chromatography as well as capillary electrophoresis coupled with mass spectral analysis are the candidate separation techniques for some cases [5]. In addition, utilization of Electro-Spray Ionization Fourier-Transform Ion Cyclotron Resonance Mass Spectroscopy (ESI-FTICR-MS) would be necessary to determine the library composition and distribution [6].

The immobilization of templates to the solid supports can facilitate the detection step of analysis in some cases. This has been successfully exploited for selective recognition of an individual enantiomer [7,8].

The selection of best-fitted receptor to a given template is a feature of DCL molecular recognition, which would lead to exploitation of an evolutionary approach to produce and separate the most appropriate receptors similar to something, which happens in the natural evolution process. The attempts to direct evolution of high-affinity ligands for biomolecules in the new emerging field called "Dynamic Diversity" can have numerous applications in self-assembly as well as drug delivery [9]. DCLs are very promising for designing enzyme inhibitors and molecular containers and molecular capsules [10].

Challenges and Limitations in Designing a DCL

There are many factors that could affect a DCL efficiency. They include (i) the nature of components and templates, (ii) the types of intermolecular interactions, (iii) thermodynamic conditions of the experiment and (iv) methods of a DCL analysis. All these factors need to be considered when designing a DCL.

(i) The nature of DCL components and templates

The components chosen are necessary to have appropriate functional groups. The more variety of these groups are present in the components the more diversity in design can be achieved (Figure 5). Furthermore, they should be selected consistent with the template properties.

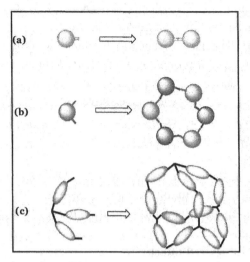

Figure 5. The number of possible aggregates would be enhanced by increasing the number of functional groups in each component and building block. a) monofunctionalized building blocks produce dimers. b) Using two functional groups macrocycles can be obtaind. c) Utilization of three functional groups would result in the capsule-like aggregates formation. A combination of the above building blocks can extend the number of possible aggregates.

(ii) The types of intermolecular interactions in DCL

It is quite useful to have prior knowledge about the intermolecular interactions between the components and the nature of component-template association so that through computational chemistry one could predict the possible nature of aggregate formation. The intermolecular interactions utilized in DCLs should be non-covalent to result in conversion reversiblility between DCL components. These interactions are expected to result in rapid equilibration so that all the available possibilities of the aggregate formation can be examined. Reversibility of reactions is responsible for institution of a dynamic exchange state in these systems and it is the paramount feature of a DCL.

(iii) Thermodynamic conditions

Solubilities of components, template and the produced aggregate in the solvent (DCL medium) could have a strong effect on the thermodynamic equilibrium conditions. To enhance the DCL efficiency, solubilities of components in the medium should not significantly differ from solubility of the template. The lack of enough template solubility is a problem, especially when using a protein as a template [3] in an aqueous medium. Utilization of nucleic acids as templates may also cause the same problem. Immobilized DNA templates have been more effective in some cases [3,11,12] than aqueous systems. Formation of an insoluble aggregate would shift the equilibrium towards the formation of that aggregate as the product.

Conditions of the reactions involved in a DCL should be as mild as possible in order to minimize the possibility of incompatibility occurrence pending exchange and molecular recognition processes.

(iv) Methods of a DCL analysis

In a DCL it should be possible to cease the ongoing reactions under certain circumstances in order to shift the system from the dynamic to the static state. Ceasing the reactions gives us the ability to switch the system off after addition of the template and formation of the best possible binders. In this situation the system is brought to an equilibration state and the distribution of aggregates are kept constant so that we can accomplish the required task. Otherwise, it is probable that changing the system conditions, while it is in the dynamic state, would change the system distribution.

An example of a DCL system with the said ability is hydrazone ($R_2C=NNR_2$) -based pseudopeptide macrocycle which uses hydrazone exchanges. A hydrazone-based library can consist of a hydrazide and protected aldehydes functionality as components [13]. The exchange and formation of hydrazone is feasible under acidic conditions and leads to formation of a library containing 10 macrocycles (from a single dipeptide building block) while neutralization of the medium causes the hydazone exchange to be ceased and switches the system off [14]. Addition of Li^+ as template induces the amplification of a cyclic trimer and decreases the concentration of other macrocyclic members to the extent that after 6 hours the said cyclic trimer becomes the only major product of the library (Figure 4). Using this method a high-affinity receptor for binding to lithium ion (Li^+) was generated which was able to change its conformation after binding to Li^+ [14].

Another similar DCL system is designed which uses oxime exchanges. The hyrazones and oximes are structurally related to the imines. Transimination of oximes in the aqueous medium are inhibited under the same conditions, namely, increase of pH from acidic to neutral and reduction of temperature will switch off the exchanges [15]. Still another similar DCL system has been designed for ammonium ion as template [16].

Before analyzing a system composition it should be ensured that the final system distribution would not be altered in the course of experimental analysis. To achieve this, it is advised to switch the library off from the dynamic to the static state during the analysis. If the type of interactions does not allow us to do so, fixation of product aggregates via covalent bonds [1,17] (for example reduction of double bonds in imines [18]) could be an alternative solution. One should be aware of a change in the final DCL distribution and the best binder concentration during such a process [3]. In short, analysis, isolation and re-utilization of the best ligand or receptor, which is formed by the molecular recognition in a DCL, are presently some of the major technical difficulties in designing such systems.

Molecular Recognition

Molecular recognition, which is the specific recognition of, and interaction with, one molecule by another, has found a great deal of applications in DCC and in host-guest chemistry [5], which is presented later in this chapter. It also has found applications to produce inclusion compounds and receptors and thus molecular containers and molecular cages through an evolutionary process.

Molding and casting are two principal concepts in molecular recognition.

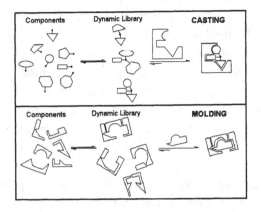

Figure 6. Symbolic demonstration of casting and molding in molecular recognition.

In "molding" a produced aggregate of a library of compounds forms a cavity within which the template is surrounded. In "casting" the space within a template acts as a cast and a location for association of a library components and entrapment of the formed aggregates. Figure 6 illustrates the principal concepts of casting and molding in molecular recognition. These concepts are also the basis of receptor formation.

A vast number of molecules have been used for self-assembly, receptor formation and molecular recognition. Such recognizing molecules could contain receptors for recognition of carboxylic acid group (-COOH) (known as carboxylic acid recognition), peptides group ($\equiv$C-N=) or carbohydrates group [19]. There are special methods and reactants for construction of cyclic, container or linear self-assembly structures (or complexes) as receptors and for molecular recognition. For instance, a strategy for cyclic-structure construction is utilization of triple and complementary hydrogen bonds between donor-donor-acceptor (DDA) group of one molecule and acceptor-acceptor-donor (AAD) group of a second molecule [20]. A common strategy for container complexes construction is using concave, bowl shaped molecular building blocks like glycoluril ($C_4H_6N_4O_2$), resorcinarenes or calixarenes as shown in Figure 7.

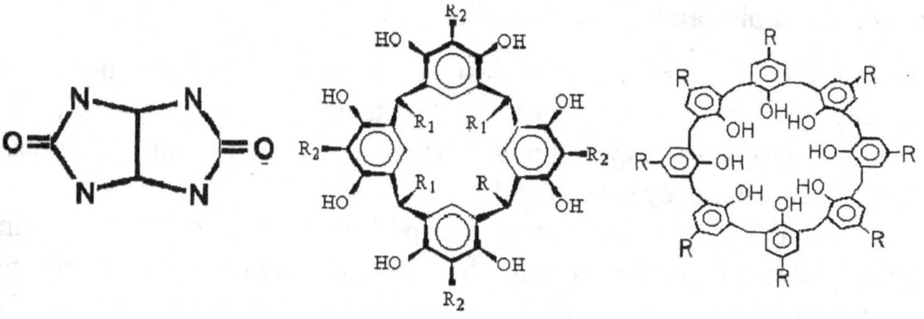

Glycoluril   Resorcinarenes   Calixarenes

Figure 7. Molecular structures of glycoluril, resorcinarenes and calixarenes used as molecular building blocks for container complexes construction.

A good example of molecular recognition casting in a DCL is utilization of carbonic anhydrase enzyme as a template in a library of imines [18]. This template could amplify the concentration of its best inhibitors among the existing aggregates in the said DCL (Figure 8).

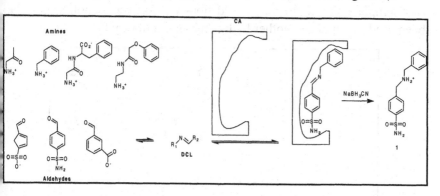

Figure 8. An imine DCL containing four amines and three aldehydes [2]. The carbonic anhydrase enzyme was added as template and the analysis of system after reduction of library by sodium cyanoborohydride indicated that the aggregate 1 has been stabilized and amplified by addition of template.

It has been shown that, by simultaneous use of an enzyme inhibitor with enzyme as template (in the control experiment), the amplifying effect will be reduced.

Some Examples and Applications of DCL

A library of peptides has been designed in order to produce the high-affinity products for an antibody (as template) [4]. However, the technical problems, especially utilization of very low concentrations, decrease its efficiency significantly.

A similar system containing tripeptides and a mixture of trypsin hydrolysates of bovine serum albumin for peptide bond exchange institution has been prepared using fibrinogen as template [2]. In this case also, the lack of enough reversibility in peptide bonds exchanges is somehow problematic.

A library of different stereoisomers was obtained using metal-ligand coordination between Fe and functionalized N-acetylgalactopyranose in which a notable aggregates concentration distribution change was observed by use of lectin [21].

Another carbohydrate-based DCL was prepared from the sugar dimers in H_2O using disulfide exchange [22].

Some efforts have been made to employ acetylcholinesterase enzyme as a template in a hydrazone-based library [23]. Acetylcholine has also been used to template another hydrazone-based library [3].

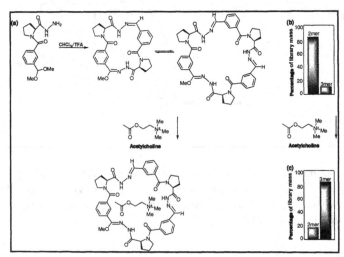

Figure 9. Effect of acetylcholine template on a hydrazone-based library [1]. Graph (b) shows the mass percentage of the dimer (red) and trimer (blue) in the absence of acetylcholine and graph (c) represents the same information for presence of the acetylcholine template condition.

In the said library a mixture of 15 macrocycles was formed by acid-catalyzed cyclization of a L-Proline-derived building block which finally changed to the mixture of two main products, a cyclic dimer (88%) and a cyclic trimer (11% as a mixture of conformers), at equilibrium [3]. Addition of acetylcholine as a template to the reaction mixture results in the significant amplification of the single conformer of the cyclic trimer that is about 50 fold over the cyclic dimer (Figure 9).

A vast number of compounds have been examined for utilization in the DCLs like polymers [24] and nucleic acids [25]. The improvement of physicochemical properties of biocompatible polymers and investigation of structure-property relationship [26] are some advantages of polymer libraries development. The pseudo-poly(amino acids) are examples of biodegradable polymers which are being developed for drug delivery systems [27]. Such is the design of a library of tyrosine derived pseudo-poly(amino acids) as reported in references [26,28].

In the context of novel receptors and catalysts design, *in vitro* evolution of the nucleic acids and design of combinatorial libraries with nucleic acid components is quite promising. The notable ability of the evolved nucleic acids and tailor-made DNAzymes in targeting and destruction of the encoding mRNA of human early growth response factor-1 (EGR-1), which is one of the contributing factors in the atherosclerosis pathogenesis, has been proved both *in vitro* and in the pig [29,30]. Moreover, targeting of the HIV-1 gag mRNA using evolved deoxyribozymes, caused reduction of HIV genes expression and inhibition of *in vitro* HIV replication [31].

It is also possible to use nucleic acids, functionalized with different functional groups (for example in the C5 position of deoxyuridine triphosphates), in the nucleic acid evolutionary systems. However, the manipulations and changes, which are made to the nucleic acid, should be tolerable by DNA and RNA polymerase enzymes and they should not disturb the base-pairing ability so that the fidelity can be preserved [25]. The utilization of such functionalized nucleic acids can increase the diversity of components and, sometimes, it can result in better aggregates with catalytic activity. For instance, the first RNA-made catalyst was evolved in a pool of modified RNA oligonucleotides, which was able to catalyze carbon-carbon bond formation in a Diels-Alder reaction [32].

Another important application of combinatorial receptor libraries is evolution and isolation of suitable catalysts [5]. Such libraries have been designed to produce and isolate the receptors which were able to catalyze the phosphodiester cleavage or cleavage of carboxylic acid esters [33].

Consequently, there is also a molecular evolution in DCL systems similar to the Darwinian evolution, which governs the nature, i.e. the selection of best-fitted and more stable species is preferred according to the physicochemical properties of a given template. The redistribution of aggregates in a DCL after addition of a template and elimination of the unfit species by the amplification of better species can be accounted as "mutation" in these systems. Hence, selection and mutation are two important features of these systems [34].

Host – Guest Chemistry

The science of host-guest chemistry was born out of a class of macrocyclic polyethers called "crown ethers" and it was first described in 1967 by C.J. Pedersen. [35].

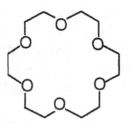

Figure 10. Molecular structure of 18-Crown-6 ether ($C_{12}H_{24}O_6$).

Presently, host-guest chemistry is referred to a branch of chemistry in which one could study the interactions between two molecules (natural or synthetic) with the goal of either mimicking or blocking an effect caused by interactions between other molecules, such as natural and biological ones [36]. For example, a number of host-guest complexes are designed to mimic the actions of enzymes. Host-guest chemistry has had a great deal of applications in biology, but its utility in molecular nanotechnology is still to be recognized. One of the leading features of the host-guest chemistry is "molecular recognition" which was presented above.

Host - guest chemistry methods can be exploited to design superamolecules, which are susceptible to recognition and specific binding to some special molecules. In these methods mostly, the inner surface of the designed molecule (host or receptor) interacts with the guest or ligand surface and weak bonds between them determine the extent of specific binding and molecular recognition.

After self-assembly, the component, which forms host, adopts an individual conformation, which often has a cavity or cleft for complete or relative entrapping of a guest molecule. Although control over the process design, and recognition specificity, in these methods are not as much as DCLs, in many cases restrictions and design difficulties are less than DCL systems. The non-covalent intermolecular interactions, which are involved in formation of such receptors, are hydrogen bonds and hydrophobic interactions. However other interactions like Br...Br, Cl...N, I...O, etc. are also used in smaller scale. Such receptors can be also synthesized on a solid support [7].

As an example of a host molecule in Figure 11 the chemical structure of calix[4]arenas is given. According to this figure calix[4]arenas are phenolic molecules which possess a bowl-shaped structure and represent a class of compounds which are subject to intensive investigations in the fields of nanotechnology self-replication and supramolecular chemistry. They contain bowl-shaped hydrophobic cavity for inclusion of various organic guest substrates. The simplest representative of these macrocycles is shown in this figure.

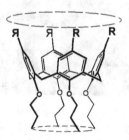

Figure 11. Chemical formula and structure of Calix[4]arenas. These macrocycle molecules possess a bowl-shaped structure.

This conformation is in dynamic (reversible) equilibrium with its inverted conformation. The chemical structure of calix[4]arene hetrodimers, which can produce molecular receptors with different shapes, are demonstrated in Figure 12.

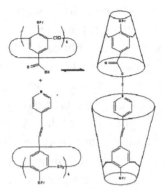

Figure 12. Calix[4]arene hetrodimers can produce molecular receptors with different shapes. (From [37]).

Introduction of groups which are capable of hydrogen bonding [20], like urea ($H_2N-C=O-NH_2$) and its derivatives, make it possible to change the shape of molecular capsule as it is demonstrated in Figures 13 and 14.

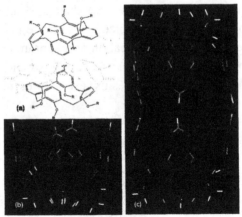

Figure 13. Introduction of hydrogen bonding groups to the calix[4]arenes to design appropriate shapes with enough stability [20].   (a) Dimeric capsule of urea ($H_2N-C=O-NH_2$) substituted calix [4] arenas . (b) Hydrogen bonding can provide essential interactions to keep bulky groups closely together and (c) coalescence of units to form a cylindrical cavity inside the dimeric capsule.

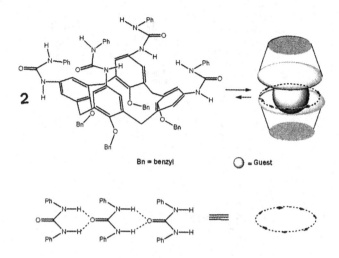

Figure 14. Urea (H_2N-C=O-NH_2) substituted calix[4]arene capsule. (From [37]).

In general, molecular capsules can be fabricated by dimerization of two concave subunits, which are able to form hydrogen bonds and encapsulate a host molecule. In principle any desired properties can be introduced to achieve an ultimate structure by using appropriate concave subunits, linkers, or spacers like an example which has been depicted in Figures 15 and 16.

Adamantane and its derivatives are widely used for kinetic study of guest release and exchanges [7,38-40]. Figure 17 shows such an exchange. Peptide nanotubes are made of chiral amino acids and can be used as host molecules for a drug (as guest molecule) delivery [41]. Besides, the adamantane constrained cyclic tripeptide; (Adm-cyst)$_3$ has been constructed which has a double-helical figure in the form of a nanotube [42]. The L-cystine can give chirality to these nanotubes while utilization of achiral compounds like adamantane makes it possible to control the size, shape and conformation of these synthetic oligopeptides [42].

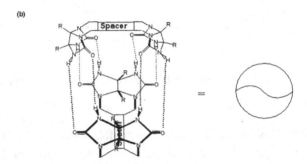

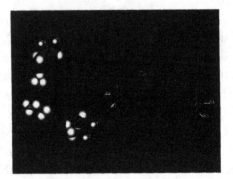

Figure 15. Utilization of concave units for self-assembly of molecular receptors [19] a) Each unit comprises of two subunits and a spacer which attaches them to each other b) Dimeric receptor as a result of hydrogen bonding between units.

Figure 16. This figure shows the concave shape of each monomer presented in Figure 1: from the side view [19]. Two monomers interact together to form a molecular container via hydrogen bonds. From top view the orientation of these two monomers seems like cross.

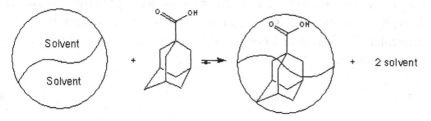

Figure 17. Two solvent molecules can be replaced by 1-adamantanecarboxylic acid. (From [19]).

Conclusions and Discussion

Combinatorial chemistry, in general, is a fast method to discovery of new materials with preferred properties. Dynamic combinatorial chemistry (DCC) is achieved when the forces of interaction between the molecules of the system are of non-covalent category. This allows reversibility in the combinatorial chemistry process, which provides flexibility and possibility of further studies with the same reacting batch. The principles of DCC are firmly established in the field of supramolecular chemistry where self-assembling systems can be used to generate molecular diversity. DCC is a potentially speedy route to new self-assemblies in nanotechnology as well as discovery of new drugs and new catalysts.

Dynamic combinatorial library (DCL) comprising of a set of intermediate compounds, in dynamic equilibrium with the building blocks, is an important step in DCC. A properly designed DCL would allow the pairing of the most suitable binder to a template.

Molecular recognition, through molding and casting, has important applications in DCC and in host-guest chemistry. Host - guest chemistry approach can be used to generate supramolecular nanostructures with desired characteristics.

The self-organizations, which occur in nature, are the origin of life producing numerous living things from single cell microorganisms to large animals and trees. Generally, the subtleties of laws of nature may not allow us to easily design compounds with preferred properties through the bottom-up self-assembly.   However, advances in DCC, molecular recognition and host-guest chemistry have paved the route to such possibilities.

Bibliography

[1]. S. Otto, R. L. E. Furlan and J. K. M. Sanders, *Curr. Opinion Chem. Bio.*, **6**, 321, (2002).

[2]. G. R. L. Cousins, S.-A. Poulsen and J. K. M. Sanders, *Curr. Opinion Chem. Bio.* **4**, 270, (2000).

[3]. S. Otto, R. L. E. Furlan and J. K. M. Sanders. *Drug Discov. Today*, **7**, 2, 117, (2002).

[4]. P. G. Swann, R. A. Casanova, A. Desai, M. M. Frauenhoff, M. Urbancic, U. Slomczynska, A. J. Hopfinger, G. C. LeBreton and D. L. Venton, *Biopolymers*, **40**, 617 (1996).

[5]. B. Linton and A. D Hamilton, *Curr. Opinion Chem. Bio.*, **3**, 307, (1999).

[6]. S.-A. Poulsen, P.J. Gates, G.R.L. Cousins and J.K.M. Sanders, *Rapid Commun. Mass Spectrom*, **14**, 44, (2000).

[7]. S. Saito and J. Rebek, *Bioorganic & Medici. Chem. Lett.*, **11**, 1497, (2001).

[8]. P. Murer, K. Lewandowski, F. Svec and J. M. J. Fréchet, *Chem. Commu.*, **23**, 2559, (1998).

[9]. C. Karan and B. L. Miller, *Drug Discov. Today*, **5**, 2, 67, (2000).

[10]. R. Fiammengo, M. Crego-Calama and D. N. Reinhoudt. *Curr. Opinion Chem. Bio.*, **5**, 660, (2001).

[11]. B. Klekota and B. J. Miller, *Tetrahedron*, **55**, 11687, (1999).

[12]. P. A. Brady and J. K. M. Sanders, *Chem. Soc. Rev.*, **26**, 326, (1997).

[13]. G. R. L. Cousins, S. A. Poulsen and J. K. M. Sanders, *Chem. Commun.*, **24**, 1575, (1999).

[14]. R. L. E. Furlan, Y. F. Ng, S. Otto and J. K. M. Sanders, *J. Am. Chem. Soc.*, **123**, 8876, (2001).

[15]. V. A. Polyakov, M. I. Nelen, N. Nazarpack-Kandlousy, A. D. Ryabov and A. V. Eliseev, *J. Phys. Org. Chem.*, **12**, 357, (1999).

[16]. R. L. E. Furlan, Y. F. Ng, G. R. L. Cousins, J. A. Redman, and J. K. M. Sanders, *Tetrahedron*, **58**, 771, (2002).

[17]. F. Cardullo, M. C. Calama, B. H. M. Snellink-Ruel, J. L. Weidmann, A. Bielejewska, R. Fokkens, N. M. N. Nibbering, P. Timmerman and D. N. Reinhoudt, *Chem. Commun.*, **25**, 367, (2000).

[18]. I. Huc and J.-M. Lehn, *Proc Natl Acad Sci USA*, **94**, 2106, (1997).

[19]. H. Chen, W. S. Weiner and A. D. Hamilton, *Curr. Opinion Chem. Bio.*, **1**, 458, (1997).

[20]. R. P. Sijbesma, E. W. Meijer, *Curr. Opinion in Colloid & Interface Sci.*, **4**, 24, (1999).

[21]. S. Sakai, Y. Shigemasa and T. Sasaki, *Tetrahedron Lett.*, **38**, 8145, (1999).

[22]. O. Ramström and J.-M. Lehn, *Chem. BioChem.*, **1**, 41, (2000).

[23]. T. Bunyapaiboonsri, *Chem. BioChem.*, **2**, 438, (2001).

[24]. S. Brocchini, *Adv. Drug Del. Rev.*, 53, 123, (2001).

[25]. J. A. Bittker, K. J. Phillips and D. R. Liu, *Curr. Opinion Chem. Bio.*, **6**, 367, (2002).

[26]. S. Brocchini, K. James, V. Tangpasuthadol, J. Kohn, *J. Biomed. Mater. Res.*, **42**, 1, 66, (1998).

[27]. K. James and J. Kohn, in *"Controlled Drug Delivery: Challenges and Strategies, Pseudo-Poly(Amino Acid)s: Examples for Synthetic Materials Derived from Natural Metabolites"*, K. Park (Ed.), Am. Chem. Soc., Washington, DC, 389, (1997).

[28]. S. Brocchini, K. James, V. Tangpasuthadol and J. Kohn, *J. Am. Chem. Soc.*, 119, **19**, 4553, (1997).

[29]. H. C. Lowe, R. G. Fahmy, M. M. Kavurma, A. Baker, C. N. Chesterman, L. M. Khachigian, *Circ. Res.*, **89**, 670, (2001).

[30]. P. O. Iversen, G. Nicolaysen and M. Sioud, *Am. J. Physiol. Heart. Circ. Physiol*, **281**, H2211, (2001).

[31]. B. Sriram and A. C. Banerjea, *Biochem J.*, **352(Pt 3)**, 667, (2000).

[32]. T. M. Tarasow, S. L. Tarasow and B. E. Eaton, *Nature*, **389**, 54, (1997).

[33]. T. Berg, A. M. Vandersteen and K. D. Janda, *Bioorg. Med. Chem. Lett.*, **8**, 1221, (1998).

[34]. B. Klekota and B. L. Miller, *TIBTECH*, **17**, 205, (1999).

[35]. C. J. Pedersen, *J. Am. Chem. Soc.*, **89**, 7017, (1967).

[36]. D. J. Cram and J. M. Cram, *"Monographs in Supramolecular Chemistry"*, J. F. Stoddart (Ed.), Royal Soc. of Chem., Cambridge, UK, (1994).

[37]. R. G. Chapman and J. C. Sherman, *Tetrahedron*, **53**, 47, 15911, (1997).

[38]. U. Lucking, D. M. Rudkevich and J. Rebek, *Tetrahedron Lett.*, **41**, 9547, (2000).

[39]. C. L. D. Gibb, H. Xi, P. A. Politzer, M. Concha and B. C. Gibb, *Tetrahedron 2002*, **58**, 673, (2002).

[40]. A. Ikeda, H. Udzu, M. Yoshimura and S. Shinkai, *Tetrahedron*, **56**, 1825, (2000).

[41]. D. Ranganathan, C. Lakshmi and I. L. Karle, *J. Am. Chem. Soc.*, **121**, 6103, (1999).

[42]. A. Nangia, *Curr. Opinion Solid State and Materials Sci.*, **5**, 115, (2001).

Chapter 11

Molecular Building Blocks — Diamondoids

"The constraints of predictability, stability, and positional control in the face of thermal vibration all favor the use of diamondoid covalent structures in future nanomachines. Present-day experience with design and computational modeling of such devices has generated useful design heuristics, including reasons for favoring diamondoid materials with many noncarbon atoms."

K. Eric Drexler

Introduction

In this chapter, at first, a general discussion about molecular building blocks for nanotechnology is presented. Then, the remaining major part of the chapter is devoted to diamondoid molecules and their role as the molecular building blocks.

Diamondoids are organic compounds with unique structures and properties. Some derivatives of diamondoids have been used as antiviral drugs for many years. Due to their flexible chemistry, their exploitations as MBBs, to design drug delivery and drug targeting are being examined.

In this chapter some methods and concepts about the role of diamondoids as MMBs in formation of nanostructures including various aspects of self-assembly are introduced. The applications of diamondoid molecules in host-guest chemistry to construct molecular receptors through self-assembly are presented. It is concluded that diamondoids are one of the best candidates for molecular building blocks in molecular nanotechnology to design nanostructures with predetermined physicochemical properties.

Molecular Building Blocks

Two different methods are envisioned for nanotechnology to build nanostructured systems, components and materials: One method is named the "top-down" approach and the other method is named the "bottom-up" approach [1]. In the top-down approach the idea is to miniaturize the macroscopic structures, components and systems towards a nanoscale of the same. In the bottom-up approach the atoms and molecules constituting the building blocks are the starting point to build the desired nanostructure.

Various illustrations are available in the literature depicting the comparison of top-down and bottom-up approaches [1]. In order to illustrate these two routes to nanotechnology Figure 1, as an example of miniaturization of lithography and nanolithography, is presented.

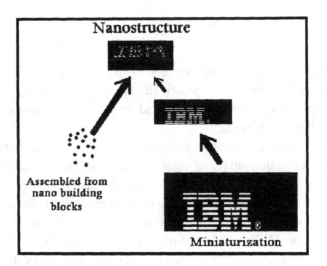

Figure 1. Schematic of top-down (miniaturization) and bottom-up approaches to nanotechnology. The nanostructure reported in this figure is a nano image [2] assembled by STM from Xenon on Nickel (110) by IBM researchers.

In the top-down method a macro-sized material is reduced in size to reach the nanoscale dimensions. The photolithography used in semiconductor industry is one example of the top-down approach. In the bottom-up strategy, we need to start with MBBs and assemble them to build a nanostructured material. The nanostructure reported in Figure 1 is the picture of a true nano image assembled by scanning tunneling microscope (STM) through the positioning of single Xenon atoms on Nickel (110) surface [2]. Since the emphasis of the present book is on the molecular based study of matter in nanoscale only the bottom-up approach is further discussed. In another words, the role of molecular building blocks to produce nanostructures will be presented.

The most fundamentally important aspect of bottom-up approach is that the nanoscale building blocks, because of their sizes of a few ϕ's [nm's], impart to the nanostructures created from them new and possibly preferred properties and characteristics heretofore unavailable in conventional materials and devices. For example, metals and ceramics produced by consolidating nanoparticles with controlled nanostructures are shown to possess properties substantially different from materials with coarse microstructures. Such differences in properties include greater hardness, higher yield strength, and ductility in ceramic materials. The band gap of nanometer-scale semiconductor structures increases as the size of the microstructure decreases, raising expectations for many possible optical and photonic applications. Considering that nanoparticles have much higher specific surface areas, thus in their assembled forms there are large areas of interfaces. One needs to know in detail not only the structures of these interfaces, but also their local chemistries and the effects of segregation and interaction among MBBs and also between MBBs and their surroundings. Knowledge on means to control nanostructure sizes, size distributions, compositions, and assemblies are important aspects of the nanoscience and nanotechnology.

The building blocks of all materials in any phase are atoms and molecules. All objects are comprised of atoms and molecules, and their arrangement and how they interact with one another define properties of an object. Nanotechnology is the engineered manipulation of atoms and molecules in a user defined and repeatable manner to build objects with certain desired properties. To achieve this goal a number of molecules are identified as the appropriate building blocks of nanotechnology. These nanosize building blocks are intermediate systems in size lying between atoms and small molecules and microscopic and macroscopic systems. These building blocks contain a limited and countable number of atoms. They can be synthesized and designed atom-by-atom. They constitute the means of our entry into new realms of nanoscience and nanotechnology.

Nanotechnology molecular building blocks (MBBs) are distinguished for their unique properties. They include, for example, graphite, fullerene molecules made of variety number of carbon atoms (C_{60}, C_{70}, etc.), carbon nanotube, diamondoids, nanowires, nanocrystals and amino acids. Figure 2 shows illustrations of fullerenes with various numbers of carbon atoms (C_{60}, C_{76}, C_{240}).

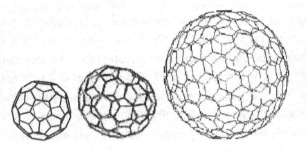

Figure 2. Illustrations of three representative fullerenes with various number of carbon atoms (C_{60}, C_{76}, C_{240}.).

In Figure 3 various forms of carbon nanotubes are reported. In Figure 4 a positional assembly taking advantage of fullerenes, carbon nanotubes, etc. as the MBBs are reported. The principles and methods of positional assembly is presented in Chapter 8 of this book. Figure 5 shows the chemical structure of amino acids and DNA. While DNA is formed as a result of nano assembly of amino acid molecules, DNAs themselves can be used as the MBBs to assemble larger nanostructures. In Figure 6 an image of self-assembled nanowires of silver with the domains between the wires made of Polymethylmethacrylate is reported. The center-to-center spacing between neighboring wires is known to be 50 nanometers.

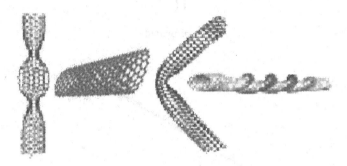

Figure 3. Various illustrations of carbon nanotube.

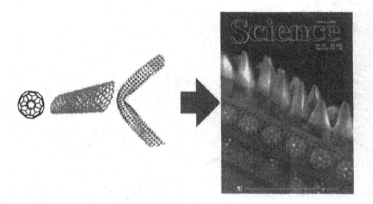

Figure 4. Positional assembly taking advantage of fullerenes, carbon nanotubes, etc. as MBBs to construct a nanostructure.

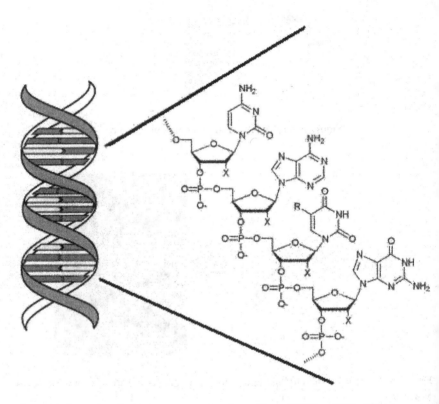

Amino acid

Figure 5. General molecular structures of amino acids and DNAs. While DNA is a result of self-assembly of amino acid molecules it also acts as the molecular building block (MBB) for larger nanostructures.

Figure 6. Self-assembled nanowires of silver. Domains between the wires are made of Polymethylmethacrylate. The center-to-center spacing between neighboring wires is 50 nanometers. Image courtesy of Heinrich Jaeger and Ward Lopes of the University of Chicago.

Figure 7 illustrates the assembly of nanocrystals of silicon atoms proposed to be used as a flash-memory chip for retaining data.

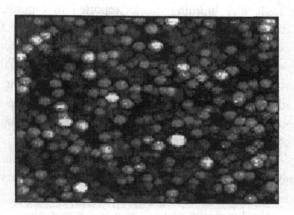

Figure 7. Illustration of nanocrystal-assembly of silicon atoms manufactured by Motorola, which are used as a flash-memory chip for retaining data.

In Figure 8 adamantane ($C_{10}H_{16}$), diamantane ($C_{14}H_{20}$), and triamantane ($C_{18}H_{24}$), the smaller diamondoid molecules, with the general chemical formula $C_{4n+6}H_{4n+12}$ are reported.

Figure 8. Molecular structures of adamantane, diamantane and trimantane, the smaller diamondoids, with chemical formulas $C_{10}H_{16}$, $C_{14}H_{20}$ and $C_{18}H_{24}$, respectively.

The lower adamantologues, including adamantane, diamantane, and triamantane, each has only one isomer. Depending on the spatial arrangement of the adamantane units, higher polymantanes can have numerous isomers and non-isomeric equivalents.

All the molecular building blocks reported in Figures 1 to 8 are candidates for various applications in nanotechnology. Diamondoid family of compounds are one of the best candidates for molecular building blocks (MBBs) to construct nanostructures compared to other MBBs known so far [3-6]. The possibility of introducing six linking groups to adamantane and thus being able to form three-dimensional structures out of connecting several adamantanes is its major potential feature which has attracted a great deal of attention.

The above mentioned building blocks have quite unique properties which are not found in rather small molecules. Some of these MBBs are electrical conductors, some are semiconductors, some are photonic and the characteristic dimension of each is a few nanometers. For example, carbon nanotubes are about five times lighter and five times stronger than steel. Many nanocrystals are photonic and they guide light through air since their spacing of the crystal pattern is much smaller than the wavelength of light being controlled. Nanowires can be made of metals, semiconductors, or even different types of semiconductors within a single wire. They're upwards of ten nanometers and they can be made to be conductor or semiconductor. Amino acids and DNA, the basis for life, can also be used to build nanomachines. Adamantane (a diamondoid) is a tetrahedrally symmetric stiff hydrocarbon that provides an excellent building block for positional (or robotic) assembly as well as for self-assembly. In fact, over 20,000 variants of adamantane have been identified and synthesized and even more are possible [4], providing a rich and well-studied set of MBBs.

It should be pointed out that MBBs with three linking groups, like graphite, could only produce planar or tubular structures. MBBs with four linking groups can form three dimensional diamond lattices. MBBs with five linking groups can create 3-dimensional solids and hexagonal planes. The ultimate present possibility is MBBs with six linking groups. Adamantane and buckyballs (C_{60}), are of the latter category which can construct cubic structures [4]. Such MBBs can have many applications in nanotechnology and they are of major interest in designing shape-targeted nanostructures, nanodevices, molecular machines [3,5,6], nanorobots and synthesis of supramolecules with manipulated architectures. As an example, the possibility of using diamondoids in designing an artificial red blood cell, called Respirocyte, which has the ability to transfer respiratory gases and glucose has been studied [7-9].

The applications of molecular building blocks (MBBs) would enable the practitioner of nanotechnology to design and build systems on a nanometer scale. The controlled synthesis of MBBs and their subsequent assembly (self-assembly, self-replication or positional-assembly) into nanostructures is one fundamental theme of nanotechnology. These promising nanotechnology concepts with far reaching implications (from mechanical to chemical processes, from electronic components to ultra-sensitive sensors; from medical applications to energy systems, and from pharmaceutical to agricultural and food chain) will impact every aspect of our future.

Diamondoids

Some Physical and Chemical Properties of Diamondoid Molecules:
Diamondoid molecules are cage-like saturated hydrocarbons. These molecules are ringed compounds which have a diamond-like structure consisting of a number of six-member carbon rings fused together. They are called "diamondoid" because they can be assumed as repeating units of the diamond. The most famous member of this group, Admantane, is a tricyclic saturated hydrocarbon (tricyclo *[3.3.1.1]*decane). The common formula for this group is $C_{4n+6}H_{4n+12}$, where $n=1$ for admantane, $n=2$ for diamantane and so on. The first three compounds of this group do not possess isomeric forms while from n≥4 the number of isomers will significantly increase. There are three possible tetramantanes all of which are isomeric, respectively as iso-, anti- and skew-tetramantane as depicted in Figure 9.

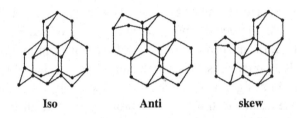

| Iso | Anti | skew |

Figure 9. There are three possible tetramantanes all of which are isomeric, respectively as iso-, anti- and skew-tetramantane. Anti- and skew-tetramantanes, each, possess two quaternary carbon atoms, whereas iso-tetramantane has three quaternary carbon atoms. The number of diamondoid isomers increase appreciably after tetramantane.

Anti- and skew-tetramantanes each possess two quaternary carbon atoms, whereas iso-tetramantane has three. There are seven possible pentamantanes, six being isomeric $(C_{26}H_{32})$ obeying the molecular formula of the homologous series and one non-isomeric $(C_{25}H_{30})$. For hexamantane, there are 24 possible structures: among them, 17 are regular cata-condensed isomers with the chemical formula $(C_{30}H_{36})$, six are irregular cata-condensed isomers with the chemical formula $(C_{29}H_{34})$, and one is peri-condensed with the chemical formula $(C_{26}H_{30})$.

When in solid state, diamondoids melt at much higher temperatures than other hydrocarbon molecules with the same number of carbon atoms in their structure. Since they also possess low strain energy they are more stable and stiff that resemble diamond in a broad sense. They contain dense, three dimensional networks of covalent bonds, formed chiefly from first and second row atoms with a valence of three or more. Many of the diamondoids possess structures rich in tetrahedrally coordinated carbon. They are materials with superior strength to weight ratio, as much as 100 to 250 times as strong as Titanium, but much lighter in weight. In addition to applications in nanotechnology they are being considered to build stronger, but lighter, rocket and other space components and a variety of other earth-bound articles for which the combination of light weight and strength is a consideration [10,11].

It has been found that adamantane crystallizes in a face-centered cubic lattice, which is extremely unusual for an organic compound. The molecule therefore should be completely free from both angle and torsional strain, making it an excellent candidate for various nanotechnology applications. At the beginning of growth, crystals of adamantane show only cubic and octahedral faces. The effects of this unusual structure upon physical properties are striking. Adamantane is one of the highest melting hydrocarbons known (m.p.~267 °C), yet it sublimes easily, even at atmospheric pressure and room temperature. Because of this, it can have interesting applications in nanotechnology.

Diamondoids can be divided into two major clusters based upon their size: lower diamondoids (*1-2* ϕ [*nm*] in diameter) and higher diamondoids (>2 ϕ [*nm*] in diameter).

Adamantane was originally discovered and isolated from Czechoslovakian petroleum in 1933. It is one of the constituents of petroleum and it deposits in natural gas and petroleum crude oil pipelines causing fouling [11,12]. The unique structure of adamantane is reflected in its highly unusual physical and chemical properties, which can have many applications in nanotechnology as do the diamond nano-sized crystals, with a number of differences. The carbon skeleton of adamantane comprises a cage structure, which may be used for encapsulation of other compounds, like drugs for drug-delivery. Because of this, adamantane and other diamondoids are commonly known as cage hydrocarbons. In a broader sense they may be described as saturated, polycyclic, cage-like hydrocarbons. The diamond-like term arises from the fact that their carbon atom structure can be superimposed upon a diamond lattice.

Diamantane, triamantane and their alkyl-substituted compounds, just as adamantane, are also present in certain petroleum crude oils. Their concentrations in crude oils are generally lower than that of adamantane and its alkyl-substituted compounds. In rare cases, tetra-, penta-, and hexamantanes are also found in petroleum crude oils. Their diamond-like rigidity and strength make them ideal for molecular building blocks. Research along this line could have a significant impact on nanomechanics such as nanomotors and nanogears. Vasquez *et al* [11,12] succeeded in identifying diamondoids in petroleum crude oils, measuring their concentrations and separating them from petroleum. Recently Dahl *et al* claim to have identified higher diamondoids, from n=4 to n=11, and their isomers as the building blocks for nanotechnology [3].

Diamondoids show unique properties due to their exceptional atomic arrangements. These compounds are chemically and thermally stable and strain-free. These characteristics make them to have a high melting point in comparison with other hydrocarbons. For instance, the *m.p.* of adamantine is estimated to be in the range of *266-268°C* and of diamantane in the range of *241-243°C*. Such high melting points of diamondoids have caused fouling in oil wells, transport pipelines and processing equipment during production, transportation and processing of diamondoids-containing petroleum crude oil and natural gas [11,12].

One may exploit the large differences in melting points of diamondoids and other petroleum fractions for isolation of diamondoids from petroleum [10,13]. Many of the diamondoids can be brought to macroscopic crystalline forms with some special properties. For example, in it's crystalline lattice, pyramidal *[1(2,3)4]*pentamantane has a large void in comparison with similar crystals. Although it has a diamond-like macroscopic structure, it possesses weak non-covalent intermolecular van der Waals attractive forces involved in forming crystalline lattice [13,14].

The crystalline structure of 1,3,5,7-tetracarboxy adamantane is formed via carboxyl hydrogen bonds of each molecule with four tetrahedral nearest-neighbors. The similar structure in 1,3,5,7-tetraiodoadamantane crystal would be formed by I...I interactions. In 1,3,5,7-tetrahydroxyadamantane, the hydrogen bonds of hydroxyl groups produce a crystalline structure similar to inorganic compounds, like CsCl, lattice [15] (see Figure 10).

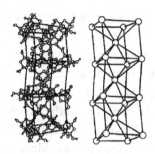

Figure 10. The quasi-cubic units of crystalline network for 1,3,5,7-tetrahydroxyada-mantane (left) is similar to CsCl (right). (From [15]).

Presence of chirality is another important feature in many derivatives of diamondoids. Such chirality among the unsubstituted diamondoids occurs first of all in tetramantane [13].

The vast number of structural isomers and streoisomers is another property of diamondoids. For instance, octamantane possesses hundreds of isomers in five molecular weight classes. The octamantane class with formula $C_{34} H_{38}$ and molecular weight 446 has 18 chiral and achiral isomeric structures. Furthermore, there are unique and great geometric diversity with these isomers. For example rod-shaped diamondoids (which the shortest one is 1.0ϕ [nm]), disc-shaped diamondoids and screw-shaped ones (with different helical pitches and diameters) have been recognized [13].

Diamondoids possess great capability for derivatization. This matter is of importance for reaching to suitable molecular geometries needed for molecular building blocks of nanotechnology. These molecules are substantially hydrophobic and their solubility in organic solvents are a function of that (admantane solubility in THF is higher than in other organic solvents [16]). Using derivatization, it is possible to alter their solubility. Functionalization by different groups can produce appropriate reactants for desired reactions.

Strain-free structures of diamondoids give them high molecular rigidity, which is quite important for a MBB. High density, low surface energy, and oxidation stability are some other preferred diamondoids properties as MBBs.

Synthesis of Diamondoids: Besides the natural gas and petroleum crude oils as the source for diamondoids one may produce them through synthesis starting with adamantane. Producing heavy diamondoids via synthetic method is not convenient because of their unique structural properties and specially their thermal stability. However, outstanding successes have been achieved in synthesis of adamantane and other lower molecular weight diamondoids. Adamantine was synthesized in 1941 for the first time [17], however, the yield was very low. Some new methods have been developed since that time and the yield has been increased to 60% [18]. The usage of zeolites as the catalyst in synthesis of adamantine has been investigated and different types of zeolites have been tested for achieving better catalyst activity and selectivity in adamantane formation reactions [17]. Recently, Masatoshi Shibuya *et al.* have represented two convenient methods for synthesis of enantiomeric *(optical antipode)* adamantane derivatives [19].

General Applications of Diamondoids: Each successive higher diamondoid family shows increasing structural complexity and varieties of molecular geometries. Sui generis properties of diamondoids have provoked an extensive range of inquiries in different fields of science and technology. For example, they have been used as templates for crystallization of zeolite catalysts [20], the synthesis of high-temperature polymers [21], and in pharmacology.

In pharmacology, two adamantane derivatives, Amantadine (1-adamantaneamine hydrochloride) and Rimantadine (α-methyl-1-

adamatane methylamine hydrochloride) have been well-known because of their antiviral activity (Figure 11).

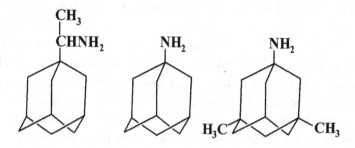

Figure 11. Amantadine (left) and Rimantadine (center) and Memantine (right), the three adamantane derivatives used as antiviral drugs.

The main indication of these drugs is prophylaxis and treatment of influenza A viral infections. They are also used in the treatment of parkinsonism and inhibition of hepatitis C virus (HCV) [22]. Memantine (1-amino-3,5- dimethyladamantane) has been reported effective in slowing the progression of Alzheimer's Disease [22].

Extensive investigations have been performed related to synthesis of new adamantane derivatives with better therapeutic actions and less adverse effects. For example, it has been proved that adamantylaminopyrimidines and -pyridines are strong stimulants of tumor necrosis factor (TNF-α) [23], or 1,6-diaminodiamantane possesses an antitumor and antibacterial activity [24]. Many derivatives of aminoadamantanes have antiviral activity and 3-(2-adamantyl) pyrolidines with two pharmacophoric amine groups have antiviral activity against influenza A virus [25].

Some derivatives of adamantane with antagonist effects have also been synthesized. For instance, monocationic and dicationic adamantane derivatives block the AMPA and NMDA receptors [26-28] and also 5-HT3 receptors [29].

Attaching some short peptidic sequences to adamantane makes it possible to design novel antagonists (Bradykinin antagonists [30] and vasopressin receptor antagonists are two examples).

Application of Diamondoids as MBBs: The importance of applications of diamondoids in nanotechnology is their potential role as MBBs. Diamondoids have noticeable electronic properties. In fact, they are H-terminated diamond and the only semiconductors which show a negative electron affinity [3].

Diamondoids can be used in self-assembly, positional assembly, nanofabrication and many other important methods in nanotechnology. Also they have found applications in drug delivery and drug targeting systems and pharmacophore-based drug design. Furthermore, they have the potential to be utilized in rational design of multifunctional drug systems and drug carriers. In host-guest chemistry and combinatorial chemistry there is plenty of room for working with diamondoids. The other potential application of diamondoids is in designing molecular capsules and cages for drug delivery.

Diamondoids, specially adamantane, it's derivatives and diamantane, can be used for improvement of thermal stability and other physicochemical properties of polymers and preparation of thermosetting resins which are stable at high temperatures. For example diethynyl diamantane has been utilized for such an application [24]. As another example, adamantyl-substituted poly(m-phenylene) is synthesized starting with 1,3-dichloro-5- (1- adamantyl) benzene monomers (Figure 12) and it is shown to have a high degree of polymerization and stability decomposing at high temperatures of around 350°C [31].

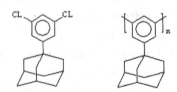

Figure 12. (Left) 1,3-dichloro-5-(1-admantyl) benzene monomer and (Right) adamantly-substituted poly(m-phenylene) which is shown to have a high degree of polymerization and stability decomposing at high temperatures of around 350°C (From [31]).

In another study, introducing adamantyl group to the poly (etherimide) structure caused polymer glass transition temperature, T_g, and solubility enhancement in some solvents like chloroform and other aprotic solvents [32].

Introducing bulky side-chains which contain adamantyl group to *PPV* elicits diminution of intra-chain interactions and thus, aggregation quenching would be reduced and polymer photoluminescence properties would be improved [33].

Substitution of the bulky adamantly group on the C(10) position of the biliverdin pigments structure, would lead to the distortion of helical conformation and hence, the pigment color would shift from blue to red [34].

Adamantane can be used in molecular studies and preparation of fluorescent molecular probes [35]. Because of its incomparable geometric structure, the adamantane core can impede interactions of fluorophore groups and self-quenching would diminish due to steric hindrance. Hence, mutual quenching would be diminished and it becomes possible to introduce several fluorescent groups to the same molecular probe in order to amplify the signals. Figure 13 shows the general scheme of an adamantane molecule with three fluorophore groups (F1) and a targeting group for attachment of biomolecules.

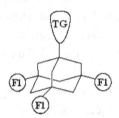

Figure 13. Schematic drawing which shows adamantane as a molecular probe with three fluorophore groups (F1) and a targeting part (TG) for specific recognition (From [36]).

Such a molecular probe can be very useful in DNA probing and especially in fluorescent-*in-situ* hybridization (FISH) diagnostics [36].

Diamondoids for Drug Delivery and Drug Targeting: Adamantane derivatives can be employed as carriers for drug delivery and targeting systems. Due to their high lipophilicity, attachment of such groups to drugs with low hydrophobicity would lead to increment of drug solubility in lipidic membranes and thus uptake increases. Furthermore, incomparable geometric properties of adamantane and other diamondoids, make it possible to introduce to them several functional groups consisting of drug, targeting part, linker, etc, without undesirable interactions. In fact, adamantane derivatives can act as a central core for such drug systems. Short peptidic sequences (like proteins and nucleic acids), lipids and polysaccharides can be bound to adamantine and provide a binding site for connection of macromolecular drugs as well as small molecules. A successful example is application of adamantyl moiety for brain delivery of drugs [37]. For this purpose, 1-adamantyl moiety was attached to several AZT (Azidothymidine) drugs via an ester spacer. These prodrugs could pass the blood brain barrier (BBB) easily. The drugs concentration after using such lipophilized prodrugs was measured in brain tissue and showed an increase of 7-18 folds in comparison with AZT drugs without adamantane vector. Ester bond would be cleaved after passing BBB by brain tissue esterases. However, the ester link should be resistant to the plasma esterases. Furthermore, because half-life of the two antiviral drugs (amantadine and rimantadine) used for this purpose (see Figure 11) in bloodstream is long (12–18 hours for amantadine and 24-36 hours for rimantadine in young adults), utilization of adamantane derivative carriers can prolong drug presence in blood circulation. However, related data for each system should be obtained. Finally, it is of importance to note that adamantane has appeared as a successful brain-directed drug carrier. Adamantane has also been used for lipidic nucleic acid synthesis as a hydrophobic group [38].

Two major problems in gene delivery are nucleic acids low uptake by cells and instability in blood medium. Probably, an increase in lipophilicity using hydrophobic groups would lead to improvement of uptake and an increase in intracellular concentration of nucleic acids [38]. For this purpose, an amide linker is used to attach adamantane derivatives to a nucleic acid sequence [38]. Such a nucleic acids derivatization has no significant effect on hybridization with target RNA.

Recently, synthesis of a polyamine adamantane derivative has been reported which has a special affinity for binding to DNA major grooves

[39]. It should be pointed out that most of polyamines have affinity for binding to RNA thus making RNA stabilized. Combined DNA and RNA binding selectivity to polyamine adamantane derivatives is one of the outstanding features of said ligand. This positive nitrogen-bearing ligand has more tendency to establish hydrophobic interactions in deeper grooves due to its size and steric properties. Such an exclusive behavior occurs because ligand fits better to DNA major grooves. This bulky ligand size is the same as zinc-finger protein. The said protein binds to DNA major grooves.

Higher affinity of adamantane-bearing ligand to DNA, instead of RNA, arises from the presence of adamantane and leads to DNA stabilization. This fact can be exploited for using such ligands as stabilizing carrier in gene delivery. Adamantine causes lipophilicity increase as well as DNA stabilization. On the other hand, a targeting sequence can be utilized in order to achieve intracellular targeting.

Ligand/groove size-based targeting is also possible with less specificity by changing the bulk and conformation of ligand. It can be used, both, for targeting of special genume regions and for better understanding of an individual region folding in the genume. Such adamantane-bearing ligands ability to establish weak specific interactions with nucleic acids helps to bring about more control over self-assembly processes. Lipidic nucleic acids possessing adamantane derivative groups can also be exploited for gene delivery. Since the nanostructures which are made from pure DNA backbone (as shown in Figure 9 in Chapter 9) do not possess enough rigidity, introduction of adamantane cores to such structures can provide adequate solidity for maintaining the desired conformation.

Polymers conjugated with 1-adamantyl moieties as lipophilic pendent groups can be utilized to design nanoparticulate drug delivery systems. For example, Polymer 1 in Figure 14, which is synthesized by homopolymerization of ethyladamantyl malolactonate, can be employed as highly hydrophobic blocks to construct polymeric drug carries [40]. As another example, polymer 2 in Figure 14, which is synthesized by copolymerization of polymer 1 with benzyl malolactonate, is water-soluble and it's lateral carboxylic acid functions can be used to bind biologically active molecules in order to perform drug-targeting as well as drug-delivery. These polymers are signified for production of pH-dependant hydrogels and intelligent polymeric systems [40].

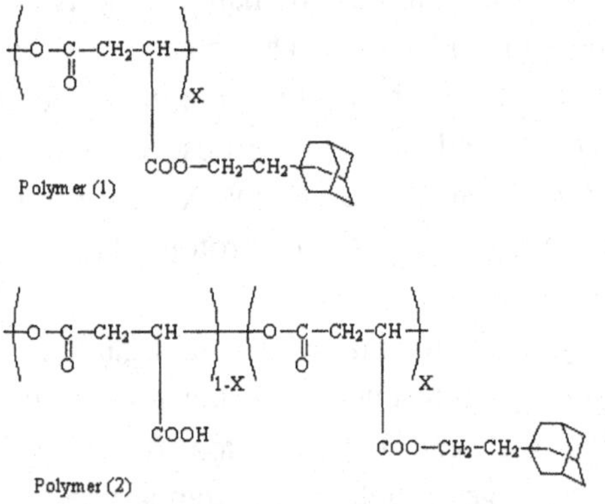

Polymer (1)

Polymer (2)

Figure 14. Polymers conjugated with 1-adamantyl moieties as lipophilic pendent groups. Polymer (1): {poly (ethyladamantyl B-malate)} can be employed as highly hydrophobic blocks to construct polymeric drug carries. Polymer (2): {poly(B-malic acid-co-ethyladamantyl B-malate)} is water-soluble and it's lateral carboxylic acid functions can be used to bind biologically active molecules in order to perform drug-targeting as well as drug-delivery. (From [40]).

DNA Directed Assembly and DNA-Adamantane-Protein Nanostructures: Due to the ability of adamantane for attachment to DNA, it seems that construction of well-defined nanostructures consisting of DNA fragments as linkers between adamantane (and it's derivatives) cores is theoretically possible and it can be a powerful tool to design nanostructured self-assemblies. The unique feature of DNA directed assembly, namely site selective immobilization, makes it possible to arrange completely defined structures. On the other hand, the possibility of introduction of vast number of substitutes, like peptidic sequences, nucleoproteins, hydrophobic hydrocarbon chains, etc to the adamantane core makes such a process capable of designing steric conformation via setting hydrophobic/ hydrophilic (and other) interactions. In addition, due to the rigidity of diamondoid structures the required strength and integrity could be provided to such self-assemblies.

Bifunctional adamantine nucleus, as a hydrophobic central core, can be used to construct peptidic scaffolding [41] as shown in Figure 15.

This fact indicates why adamantane is considered as one of the best MBBs. It seems an effective and practical strategy to substitute different amino acids on the adamantane core (Figure 16) and exploit nucleic acid sequences as linkers and DNA hybridization to attach these modules.

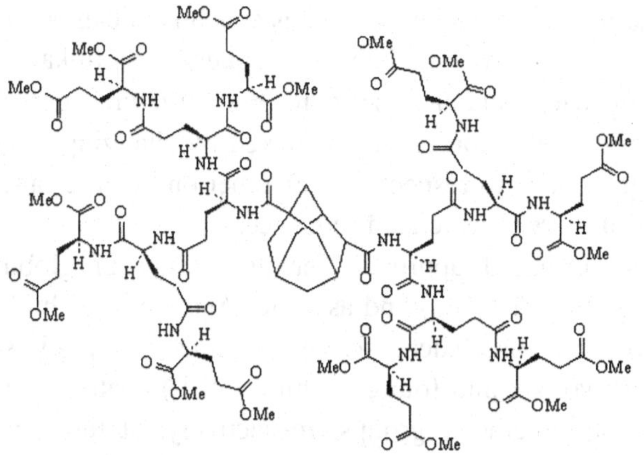

Figure 15. Adamantane nucleus with amino acid substituents creates a peptidic matrix. The represented structure is Glu_4-Glu_2-Glu-[ADM]-Glu-Glu_2-Glu_4 (From [41]).

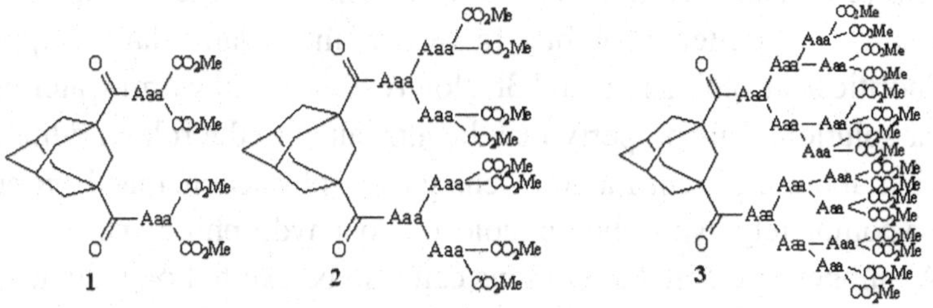

a : Aaa = Asp ; b : Aaa = Glu

Figure 16. Amino acid substituted adamantane cores can be employed as diverse nanomodules. (from [41]).

Thus, the DNA-Adamantane-Protein nanostructures can be constructed. The knowledge about proteins folding and conformations in biological systems can also inspire us to design such nanostructures with a desired and predictable conformation in a biomimetic way. Step by step assembly makes it possible to construct a desired nanostructure and examine the effect of each new MBB on the nanostructure. Such a controlled and directed assembly can be utilized to investigate molecular interactions, molecular modeling and study of relationships between the composition of MBBs and the final conformation of the nanostructures. Immobilization of molecules on a surface could facilitate such studies [42]. For adamantane cores nucleic acids attachment can be introduced in many ways. At least, two nucleic acid sequences as linkage groups are necessary for each adamantane core but structural development in desirable steric orientation can be achieved by changing the position of two said sequences with respect to each other on the adamantane core or introduction of more nucleic acid sequences.

A dendrimer-based approach for the design of globular protein mimics using glutamic (Glu) and aspartic (Asp) acids as building blocks is developed [41]. Three successive generations of peptidic scaffoldings consisting of two, six and fourteen chiral (all L) centres and four, eight and sixteen carbomethoxy groups, respectively, at the periphery with adamantane nucleus as the central core have been constructed by linking the two halves of corresponding Asp/Glu dendrous by 1,3 - bifunctional adamantane unit. An interesting feature of GluAsp dendritic scaffoldings is their remarkable solubility behavior. Thus, whereas Glu, and Glu dendrons, are quite insoluble in water the adamantane supported bisdendritic scaffoldings 2b and 3b slowly dissolve in warm water giving a clear solution. This property may be attributed to the role of admantane bringing about different range of changes in physicochemical properties from a completely hydrophobic molecule to a hydrophilic one.

A number of desired alterations can also be exerted on a nucleic acid sequence utilizing the new techniques developed in solid-phase genetic engineering for immobilized DNA alteration. For instance, DNA ligation can be employed to join nanomodules (as well as hybridization) similar to which has been done on immobilized DNA in the case of gene assembly [44]. It is possible to modify the amino acid parts of the said nanostructures as well as adamantane cores and DNA sequences. For example, using some unnatural (synthetic) amino acids [45] with appropriate folding characteristics, the ability of conformation fine-tuning could be improved. Moreover, polypeptides and nucleic acids are the major components for self-replication [46] and they could facilitate designing of self-replication processes for the said nanostructures. Hence, assembling and composing of adamantane nucleuses as central cores, DNA sequences as linkers and amino acid substituents (on the adamantane) as conformation controllers would lead to design of DNA-adamantane-protein nanostructures with desired and predictable properties.

Diamondoids for Host-Guest Chemistry: The paramount aim in host-guest chemistry is to construct molecular receptors by self-assembly process so that such receptors could, to some extent, gain molecular recognition capability. Calixarenes, which are macrocyclic compounds, are some of the best building blocks to design molecular hosts in supramolecular chemistry. In 2002, the first synthesis of Calix[4]arenes, which had been adamantylated on its upper rim, was reported [47]. In these Calix[4]arenes, adamantane or its ester/carboxylic acid derivatives, were introduced as substituents. The purpose of this synthesis was to learn how to employ the flexible chemistry of adamantane in order to construct different kinds of molecular hosts.

An important strategy to design host molecules is joining several concave monomers/units via a linker like Calix[4]arenas as shown by Figure 11 in Chapter 10. These monomers /units are capable of hydrogen bonding. Similarly, the synthesis of tetrameric 1,3-adamantane and it's butyl derivative has been reported [48] (see Figure 17).

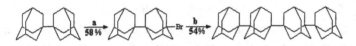

Figure 17. Synthesis of adamantane tetramers. Reagents and conditions are: (a) Br$_2$, CCl$_4$, 30°C, 72 h; (b) Na, n-octane, reflux, 12 h (From [48]).

Introduction of some groups which are capable of hydrogen bonding (like urea, amines, hydroxyl groups, etc.) to the said tetrameric 1,3-adamantane derivative could probably be exploited to construct host molecules. Calix[4]arenes are also utilized to design DCLs [49] and perhaps their adamantane derivatives can be employed in the same process for self-assembly and producing molecular receptors.

Some other types of macrocycles have been synthesized using adamantane and its derivatives. Recently, a new class of cyclobisamides has been synthesized using adamantane derivatives, which show the general profile of amino acids (serine or cystine)-ether composites, and they were shown to be efficient ion transporters (especially for Na^+ ions) in the model membranes [50].

Other interesting compounds to which adamantane derivatives have been introduced in order to obtain cyclic frameworks are "crown ethers" [51]. The outstanding feature of these adamantane-bearing crown ethers (which are also called "Diamond Crowns") is that α-amino acids can be incorporated to the adamantano-crown backbone [51]. This family of compounds provides valuable models to study selective host-guest chemistry, ion-transports and ion-complexation [51].

Conclusions and Discussion

Diamondoids are organic compounds with unique structures and properties. This family of compounds with over 20,000 variants is one of the best candidates for molecular building blocks (MBBs) to construct nanostructures compared to other MBBs known so far

Some of their derivatives have been used as antiviral drugs for many years. Due to their flexible chemistry, they can be exploited to design drug delivery systems and also in molecular nanotechnology. In such systems, they can act as a central lipophilic core and different parts like targeting segments, linkers, spacers, or therapeutic agents can be attached to the said central nucleus. Their Central core can be functionalized by peptidic and nucleic acid sequences and also numerous important biomolecules.

Furthermore, some adamantane derivatives possess special affinity to bind to DNA, making it stabilized. This is an essential feature for a gene vector. Some polymers have been synthesized using adamantane derivatives application of which is under investigation for drug delivery.

Adamantane can be used to construct peptidic scaffolding and synthesis of artificial proteins. Introduction of amino acid-functionalized adamantane to the DNA nanostructures would lead to construction of DNA-adamantane-protein nanostructures with desirable stiffness and integrity. Diamondoids can be employed to construct molecular cages and containers and also for utilization in different methods of self-assembly. In fact, by development of self-assembly approaches and utilization of diamondoids in these processes, it would be possible to construct novel nanostructures especially to design effective and specific carriers for each drug.

The phase transition boundaries (phase envelop) of adamantane need to be investigated and constructed. Predictable and diverse geometries are important features for molecular self-assembly and pharmacophore-based drug design. Incorporation of higher diamondoids in solid-state systems and polymers should provide high-temperature stability, a property already found for polymers synthesized from lower diamondoids.

Diamondoids offer the possibility of producing a variety of nanostructural shapes including molecular-scale components of machinery such as rotors, propellers, ratches, gears, toothed cogs, etc. We expect them to have the potential for even more possibilities for applications in molding and cavity formation characteristics due to their organic nature and their sublimation properties. The diverse geometries and varieties of attachment sites among higher diamondoids provide an extraordinary potential for the production of shape-derivatives.

Bibliography

[1]. R. W. Siegel, E. Hu, and M. C. Roco (Editors), *"Nanostructure Sci. and Technology - A Worldwide Study. Prepared under the guidance of the IWGN, NSTC"* WTEC, Loyola College in Maryland, (1999).

[2]. D. M. Eiger and E. K. Schweizer, *Nature*, **344**, 524, (1990).

[3]. J. E. Dahl, S. G. Liu, R. M. K. Carlson, *Science*, **299**, 96, (2003).

[4]. R. C. Merkle, *"Molecular building blocks and development strategies for molecular nanotechnology"*, at: [http://www.zyvex.com/nanotech/mbb/mbb.html].

[5]. K. E. Drexler, *TIBTECH*, **17**, 5 (1999).

[6]. H. Ramezani and G. A. Mansoori, *"Diamondoids as Molecular Building Blocks for Nanotechnology"* (to be published).

[7]. K. Bogunia-Kubik and M. Sugisaka, *BioSystems*, **65**, 123, (2002).

[8]. R. A. Freitas, *Nanotechnology*, **2**, 8, (1996).

[9]. R. A. Freitas, *Artif. Cells, Blood Subtit. and Immobil. Biotech*, **26**, 411, (1998).

[10]. L. D. Rollmann, L. A. Green, R. A. Bradway and H. K. C. Timken, *Catalysis Today*, **31**, 163, (1996).

[11]. D. Vazquez Gurrola, J. Escobedo and G. A. Mansoori, *"Characterization of Crude Oils from Southern Mexican Oilfields"* Proceedings of the EXITEP 98, International Petroleum Technology Exhibition, Placio de Los Deportes, 15th-18th November, Mexico City, Mexico, D.F., (1998).

[12]. D. Vasquez and G. A. Mansoori, *J. Petrol. Sci. & Eng.*, **26**, 1-4, 49, (2000).

[13]. G. A. Mansoori, *"Advances in atomic & molecular nanotechnology"* in *Nanotechnology, United Nations Tech Monitor*, 53, Sep-Oct (2002).

[14]. G. A. Mansoori, L. Assoufid, T. F. George and G. Zhang, "Measurement, Simulation and Prediction of Intermolecular Interactions and Structural Characterization of Organic Nanostructures" in Proceed. of Conference on Nanodevices and Systems, Nanotech 2003, San Francisco, CA, February 23-27, (2003).

[15]. G. R. Desiraju, *J. of Mol. Struc.* 1996, **374**, 191, (1996).

[16]. J. Reiser, E. Mc Gregor, J. Jons, R. Erick and G. Holder, *Fluid Phase Equil.*, **117**, 160, (1996).

[17]. M. Navratilova and K. Sporka, *Appl. Catalysis A: Gen.*, **203**, 127, (2000).

[18]. H. Hopf, *"Classics in Hydrocarbon Chemistry: Syntheses, Concepts, Perspectives"* Wiley-VCH Verlog GmbH, Weinheim, Germany (2000).

[19]. M. Shibuya, T. Taniguchi, M. Takahashi and K. Ogasawara, *Tetrahedron Lett.*, **43**, 4145, (2002).

[20]. S. I. Zones, Y. Nakagawa, G. S. Lee, C. Y. Chen, and L. T. Yuen, *Microporous and Mesoporous Mater.*, **21**, 199, (1998).

[21]. M. A. Meador, *Annu. Rev. Mater. Sci.*, **28**, 599, (1998).

[22]. J. G. Hardman, L. E. Limbird, A. G. Gilman, *"Goodman & Gilman's: The pharmacological basis of therapeutics"* 10th Ed., McGraw-Hill Pub. Co., New York, NY, (2001).

[23]. Z. Kazimierczuk, A. Gorska, T. Świtaj and W. Lasek, *Bioorg. & Medici.Chem. Lett.*, **11**, 1197, (2001).

[24]. Y.-T. Chern and J.-J. Wang, *Tetrahedron Lett.* 1995, **36**, 32, 5805, (1995).

[25]. G. Stamatiou, A. Kolocouris, N. Kolocouris, G. Fytas, G. B. Foscolos, J. Neyts and E. De Clercq, *Bioorg. & Medici.Chem. Lett.*, **11**, 2137, (2001).

[26]. K. V. Bolshakov, D. B. Tikhonov, V. E. Gmiro and L. G. Magazanik, *NeuroSci. Lett.*, **291**, 101, (2000).

[27]. M. V. Samoilova, S. L. Buldakova, V. S. Vorobjev, I. N. Sharonova and L. G. Magazanik, *NeuroSci.*, **94**, 1, 261, (1999).

[28]. S. L. Buldakova, V. S. Vorobjev, I. N. Sharonova, M. V. Samoilova and L. G. Magazanik, *Brain Research* **846**, 52, (1999).

[29]. G. Rammes, R. Rupprecht, U. Ferrari, W. Zieglgänsberger and C. G. Parsons, *NeuroSci. Lett.*, **306**, 81, (2001).

[30]. S. Reissmann, F. Pineda , G. Vietinghoff , H. Werner, L. Gera , J. M. Stewart and I. Paegelow, *Peptides*, **21**, 527, (2000).

[31]. L. J. Mathias and G. L. Tullos. *Polymer*, **37**, 16, 3771, (1996).

[32]. G. C. Eastmond, M. Gibas and J. Paprotny, *Eur. Polymer J.*, **35**, 2097, (1999).

[33]. Y. K. Lee, H. Y. Jeong, K. M. Kim, J. C. Kim, H. Y. Choi, Y. D. Kwon, D. J. Choo, Y. R. Jang, K. H. Yoo, J. Jang and A. Talaie, *Curr. Appl. Physics*, **2**, 241, (2002).

[34]. A. K. Kar and D. A. Lightner, *Tetrahedron: Asymmetry*, **9**, 3863, (1998).

[35]. P. R. Seidl and K. Z. Leal, *J. Mol. Struc. (Theochem)*, **539**, 159, (2001).

[36]. V. V. Martin, I. S. Alferiev and A. L. Weis, *Tetrahedron Lett.* 1999, **40**, 223, (1999).

[37]. N. Tsuzuki, T. Hama, M. Kawada, A. Hasui, R. Konishi, S. Shiwa, Y. Ochi, S. Futaki and K. Kitagawa, *J. Pharmaceutical Sci.*, **83**, 4, 481, (1994).

[38]. M. Manoharan, K. L. Tivel and P. D. Cook, *Tetrahedron Lett.*, **36**, 21, 3651, (1995).

[39]. N. Lomadze and H.-J. Schneider. *Tetrahedron Lett.*, **43**, 4403, (2002).

[40]. L. Moine, S. Cammas, C. Amiel, P. Guerin and B. Sebille, *Polymer*, **38**, 12, 3121, (1997).

[41]. D. Ranganathan and S. Kurur, *Tetrahedron Lett.*, **38**, 7, 1265, (1997).

[42]. K. Busch and R. Tampe, *Rev. Mol. Biotech.*, **82**, 3, (2001).

[43]. R. C. Merkle, *TIBTECH*, **17**, 271, (1999).

[44]. J. H. Kim, J.-A Hong, M. Yoon, M. Y. Yoon, H.-S. Jeong and H. J. Hwang, *J. of Biotech.*, **96**, 213, (2002).

[45]. G. Fahy, *Foresight Update 16*, Available at: [http://www.foresight.org/Updates/Update16/Update16.1.html#anchor576239], (1993).

[46]. D. H Lee, K. Severin and M. R. Ghadiri, *Curr. Opin. in Chem. Bio.*, **1**, 491, (1997).

[47]. E. Shokova, V. Tafeenko and V. Kovalev, *Tetrahedron Lett.*, **43**, 5153, (2002).

[48]. T. Ishizone, H. Tajima, S. Matsuoka and S. Nakahama, *Tetrahedron Lett.*, **42**, 8645, (2001).

[49]. M. Crego-Calama, P. Timmerman and D. N. Reinhoudt, *Angew Chem. Int. Ed. Engl.*, **39**, 755, (2000).

[50]. D. Ranganathan, M. P. Samant, R. Nagaraj and E. Bikshapathy, *Tetrahedron Lett.*, **43**, 5145, (2002).

[51]. D. Ranganathan, V. Haridas and I. L. Karle, *Tetrahedron*, **55**, 6643, (1999).

[52]. H. Chen, W. S Weiner and A. D Hamilton, *Curr. Opin. in Chem. Bio.*, **1**, 458, (1997).

Glossary

5-HT = 5-Hydroxytryptamine.
Known as serotonin, an important neurotransmitter made by neurons in the central nervous system. After it is released by a neuron, activates receptors located on nearby neurons.

ab initio [= From the beginning (Latin)] *calculation (or method)*
Calculated (formulated) without empiricism or the use of experimental data except for fundamental physical constants such as Boltzmann constant or Planck constants.. Molecular orbital calculations which use all the molecular orbitals in a calculation.

AAD = Acceptor-Acceptor-Donor

Adamantane (See diamondoid)

Adatom
Adsorbed atom on a smooth surface

Avidin
A basic glycoprotein of known carbohydrate and amino acid content. It combines stoichiometrically with biotin. The biotin-avidin complex is dependent upon media ionic strength, but relatively stable over a wide pH range and to heat. It is necessary to effect a reversible or irreversible denaturation to free biotin.

AMPA = α-amino-3-hydroxy-5-Methylisoxazole-4-Propionic Acid

Amphipathic molecule
A molecule, having both hydrophobic and hydrophilic regions. Can apply equally to small molecules, such as phospholipids and macromolecules such as proteins.

307

Amphiphile

A compound containing a large organic cation or anion which possesses a long unbranched hydrocarbon chain. The existence of distinct polar (hydrophilic) and non-polar (hydrophobic) regions in the molecule promotes the formation of micelles.

AFM =Atomic force microscope.

A tip-based piso-scanning instrument able to image surfaces to molecular accuracy by mechanically probing their surface contours. It can be used for analyzing the material surface all the way down to the atoms and molecules level. A combination of mechanical and electronic probe is used in AFM to magnify surfaces up to 100,000,000 times to produce 3-D images of them.

Antagonist effect

Effect of a drug that neutralizes or counteracts the effects of another drug.

Antibody

A special protein produced by the cells of the immune system of humans and higher animals in response to the presence of a specific antigen. It recognizes and helps fight infectious agents and other foreign substances (antigens) that invade the body by binding to them.

Apoptosis

Programmed cell death as signaled by the nuclei in normally functioning human and animal cells when age or state of cell health and condition dictates.

Aprotic Solvent

Solvents whose hydrogen atoms are not bonded to an oxygen or nitrogen atom. Examples of polar aprotic solvents include aldehydes, R-CHO, ketones, R-CO-R', Dimethyl Sulfoxide(DMSO), CH_3-SO-CH_3, Dimethyl Formamide(DMF), H-CO-N(CH_3)$_2$. Examples of non-polar aprotic solvents include all the hydrocarbons (alkanes, alkenes, and alkynes).

AZT = Azidothymidine.

The prescription drug most commonly used to treat AIDS. Since the production of new red blood cells is affected, other combination treatments are used to prevent anemia.

Base-Pairing
The hydrogen bonding of complementary nitrogenous bases, one purine and one pyrimidine, in DNA and in hybrid molecules, join DNA and RNA.

Beta-sheet = β sheet = β-pleated sheet
A commonly occurring form of regular secondary structure in proteins, as proposed by Linus Pauling and Robert Corey in 1951. β sheet consists of two or more amino acid sequences within the same protein. They are arranged adjacently and in parallel, however with alternating orientations. This allows hydrogen bonds to form between the two strands.

Blood-Brain Barrier
Physiological barrier protecting the brain tissue from toxins in the bloodstream. This protection system inhibits also the entry of most therapeutical drugs into the brain.

BCC = body-centered cubic crystalline structure.
A structure in which the simplest repeating unit consists of nine equivalent lattice points, eight of which are at the corners of a cube and the ninth of which is in the center of the body of the cube.

BIO-STV =Biotin-Straptavidin

Biomimetics
Making artificial products that mimic the natural ones based on the study of structure and function of biological substances.

Biotin
A small molecule that binds with high affinity to avidin and streptavidin. Biotin is used to label nucleic acids and proteins that may be subsequently detected by avidin or streptavidin linked to a fluorescent or enzymatic reporter molecule.

Born-Oppenheimer Approximation
The assumption that the electronic motion and the nuclear motion in molecules can be separated. It leads to a molecular wave function in terms of electron positions and nuclear positions.

Bottom-Up
Building larger objects from molecular building blocks.

Brillouin Zone
In the propagation of a wave motion through a crystal lattice, the wave frequency is a periodic function of its wave vector k. This function may be complicated by being multivalued and with discontinuities. To simplify the treatment, a zone (known as Brillouin zone) in k-space is defined. It forms the fundamental periodic zone such that the frequency or energy for a k outside this zone may be determined from one of those in it. It is usually possible to restrict attention to k values inside the Brillouin zone. Discontinuities occur only on the boundaries. If the zone is repeated indefinitely, all k-space will be filled.

Buckminsterfullerene (See also Fullerene)
The official name of C_{60} fullerene discovered in 1985. Named after the famous architect Buckminster Fuller, who proposed the geodesic dome, which fullerenes resemble.

Bucky Balls
A common name for C_{60} molecule - made up of 60 carbon atoms, arranged in a soccer ball shape forming hexagonals.

Bulk Modulus
It is equal to $-V(\partial P/\partial V)_T$, the reciprocal of the coefficient of isothermal compressibility.

CA
Carbonic anhydrase - enzymes that catalyze the hydration of carbon dioxide $(CO_2+H_2O \Leftrightarrow HCO_3^-+H^+)$ and the dehydration of bicarbonate.

Carbon Nanotube
A molecule first discovered in 1991 Sumio Iijima, made from carbon atoms connected into a tube as small as 1 ϕ [nm] in diameter. It is equivalent to a flat graphene sheet rolled into a tube with high strength capacity and lightweight.

Cauchy Relation
Cauchy relation is a simple test for deciding whether the total intermolecular potential energy of a matter can be described as a sum of pair potentials. According to the 19th century French scientist, Jean Claude Saint-Venant, the assumption of central pairwise forces proposed by Cauchy and Poisson implied a reduction in the number of independent elastic constants from 21 to 15. The remaining six constants could be

expressed by the Cauchy relation $C_{12} = C_{44}$ if atoms are at centers of cubic symmetry. For further information see: A. E. H. Love, Mathematical Theory of Elasticity, 4th ed. (Cambridge Univ. Press, 1927).

Chemset
A collection of two or more building blocks members.

Chimeric
Being or relating to or like a chimera (Greek mythology) fire-breathing she-monster with a lion's head and a goat's body and a serpent's tail; daughter of Typhon.

Chirality
A chiral molecule is that it is not super-imposable on its mirror image.

Classically Forbidden Region
According to quantum mechanics, whenever a particle attempts to enter a region of the phase space where its kinetic energy is less than the potential energy, the wavefunction will decay exponentially with distance and the particle is said to be in the "classically forbidden" region.

Coacervate
Colloidal (micelles) aggregation. One theory of the evolution of life is that the formation of coacervates in the primeval soup was a step towards the development of cells.

Combinatorial Chemistry
Using a combinatorial process to prepare sets of compounds from sets of building blocks.

Cooperative Bonding
When binding a ligand, the affinity of the ligand for the molecule changes depending on the amount of ligand already bound. A macromolecule is said to have positive cooperativity if the binding of ligand increases affinity for the ligand, and negative cooperativity if the affinity for the ligand decreases as more ligand is bound.

CSR
Compartmentalized Self-Replication.

Cystine
The product of an oxidation between the thiol side chains of two cysteine amino acids. As such, cystine is not considered one of the 20 amino acids. This oxidation product is found in abundance in a variety of proteins such as hair keratin, insulin, etc.

CNS   = Central Nervous System

Coacervation
Raised into a pile; collected into a crowd; heaped, formation of colloidal aggregates containing a mixture of organic compounds. A form of self-assembly.

Covalent Bond
The bond in compounds that results from the sharing of one or more pairs of electrons.

CSR   = Compartmentalized Self-Replication

Curie Temperature
Temperature above which a ferromagnetic material loses its magnetic properties and becomes paramagnetic.

Cyclodextrins
A homologous group of cyclic glucans consisting of alpha-1,4 bound glucose units obtained by the action of cyclodextrin glucanotransferase on starch or similar substrates. The enzyme is produced by certain species of bacillus. Cyclodextrins form inclusion complexes with a wide variety of substances

Cystine
An amino acid containing one disulphide bond found in proteins. It is produced when two cysteine molecules linked by a disulfide (S-S) bond.

Cytoplasm
The viscid, semi-fluid matter contained within the plasma membrane. A component of every biological cell, the cell membrane (or plasma membrane) is a thin and structured layer of lipid and protein molecules, which surrounds the cell. It separates a cell's interior from its surroundings and controls what moves in and out. Cell surface membranes often contain receptor proteins and cell adhesion proteins.

These membrane proteins are important for the regulation of cell behavior and the organization of cells in tissues.

Cytotoxic
Chemicals that are directly toxic to cells, preventing their reproduction or growth.

DCL = Dynamic Combinatorial Library

DDA = Donor-Donor-Acceptor

DDI = DNA-Directed Immobilization

deBroglie (Thermal) Wavelength
In wave mechanics, the de Broglie wavelength equation is $\lambda=h/p=h/(mv)$ where p is the momentum, m is the mass and v is the velocity. In 1924, de Broglie postulated a connection between radiation of material particles and photons, that wave-particle behavior of radiation applies to matter. de Broglie discovered that all particles with momentum have a wavelength, now called the de Broglie thermal wavelength, $\Lambda = [h^2/2\pi mk_B T]^{1/2}$, where h is the Planck, k_B is the Boltzmann constant, m is the particle's rest mass, and T is the absolute temperature. Λ is the mean quantum mechanical wavelength of the molecules at temperature T. In other words, it expresses the momentum term in terms of thermal energy.

Dendrimer
A polymer that branches in its growth like a tree. From the Greek word dendra – meaning tree.

Detailed Balance
In an stochastic process the detailed balance means that the balance between the probability of `leaving' a state and arriving in it from another state holds not just overall, but in fact individually for any pair of states (leaving and going to and vice versa). Important in describing the properties of the equilibrium state.

DFT =Density Functional Theory
A first principles technique to solve the Schrödinger equation. A general approach to the *ab initio* description of quantum many-particle systems, in which the original many-body problem is rigorously recast in the form of an auxiliary single-particle problem. It was developed by Walter Kohn

and co-workers in the 1960's based on the original ideas of Thomas and Fermi.

Diamondoid
Cage-like saturated hydrocarbons with strong stiff structures resembling diamond. Contain three-dimensional networks of covalent bonds. Considered the favorite molecular building blocks for nanotechnology.

Diatom Frustules
The silica-rich cell wall of a microscopic, single-celled algae called diatom. Diatoms are so abundant that they can form thick layers of sediment composed of the frustules of the organisms that died and sank to the bottom. Frustules have been an important component of deep-sea deposits since Cretaceous time.

Diels-Alder Reaction
A cycloaddition reaction in which an alkene reacts with a 1,3-diene to form a 6-membered cyclic ring.

Disproportionation
Oxidation-reduction reaction in which the same element is both oxidized and reduced.

DNA = Deoxyribonucleic Acid
The carrier of genetic information, which passes from generation to generation. Every cell in the body, except red blood cells, contains a copy of the DNA.

DNA Ligation
Joining linear DNA fragments together with covalent bonds

DNA-MARKER =Y-DNA
Unique DNA sequences (The ordered arrangement of the bases within DNA) used to characterize or keep track of a gene, chromosome or DNA lineage.

DNAzyme
Association of a DNA and an enzyme.

Dry Nanotechnology
Non-biological and inorganic nanotechnology focusing on fabrication of structures in carbon, silicon, and other inorganic materials including

metals, semiconductors, etc. Developing solid-state electronic, magnetic, and optical devices and structures.

Dynamic Library

Collection of compounds in dynamic equilibrium. If the composition of the library is altered, for instance by the presence of a receptor which selectively binds certain library members, then shifting of the equilibrium will lead to an increase in the amount of those components which bind to the target with relatively high affinity.

EAM = Embedded-atom model potential.

This is a type of interatomic potential, first suggested by Friedel, which describes the bonding of an atom in terms of the local electronic density.

Eigenvalue; Eigenvector

If a (scalar) value, s, satisfies $Mv = sv$ for some non-zero vector, v, it is an eigenvalue of the matrix M, and v is an eigenvector. In mathematical programming this arises in the context of convergence analysis, where M is the hessian of some merit function, such as the objective or Lagrangian. Matrices, which are not symmetric, have right eigenvectors and left eigenvectors, even though the spectrum is the same. Right eigenvector means $Mv=sv$ where M is the matrix and s is its eigenvalue. Left eigenvector means $^T vM=v^T m$.

Elastic Moduli

A set of constants, also known as elastic constants, that defines the properties of material that undergoes stress, deforms, and then recovers and returns to its original shape after the stress ceases. They include the bulk modulus, Lame constant, Poisson's ratio, shear modulus, and Young's modulus.

Enantiomer

Either of a pair of chemical compounds (isomers) whose molecular structures have a mirror-image relationship to each other called also optical antipode

Endocytosis

Uptake of material into a cell by the formation of a membrane bound vesicle.

Endosome
Endocytotic vesicle derived from the plasma membrane. More specifically an acidic nonlysosomal compartment in which receptor ligand complexes dissociate.

Entropy
A property of a system with dimension of energy divided by temperature. The second law of thermodynamics states that the entropy production in every non-equilibrium system is always positive and it is zero for systems at equilibrium.

ERG =Early Growth Response factor

Ergodicity
A stochastic system that tends in probability to a limiting form that is independent of the initial conditions is called ergodic. The interchangeability of the time average with the ensemble average in statistical mechanics is called ergodicity.

Esterases
Any of various enzymes that catalyze the hydrolysis of an ester.

Ewald Sum Method
An efficient technique for treating long-range interactions (like electrostatics) between particles and all their infinite periodic images. It is written as the sum of three terms, namely, the direct space term ϕ^D, the Fourier term ϕ^F and the constant term ϕ^C as shown below [P. Ewald, *Ann. Phys.* **64**, 253, (1921)]:

$$\phi = \phi^D + \phi^F + \phi^C$$

$$\phi^D = \left(\frac{1}{2}\right)\sum_{i,j}^{N'}\sum_n \ q_i q_j \frac{erfc(\alpha r_{ij,n})}{r_{ij,n}} \ ; \ \ erfc(x) = 1 - \frac{2}{\sqrt{\pi}}\int_0^x \exp(-t^2)dt$$

$$\phi^F = \left(\frac{1}{2\pi}\right)\sum_{i,j}^{N}\sum_n \ q_i q_j \sum_{m \neq 0}^{N} \frac{\exp[-(\pi m/\alpha)^2 + 2\pi i m(\vec{r}_i - \vec{r}_j)]}{m^2},$$

$$\phi^C = \left(\frac{\alpha}{\sqrt{\pi}}\right)\sum_i^{N'} \ q_i^2$$

In principle Ewald sum is exact and converges much more rapidly than the regular electrostatic potential function.

Excretory Mechanism
Bodily process for disposing of undigested food waste products and nitrogenous by-products of metabolism; regulating water content, maintaining acid-base balance, and controlling osmotic pressure to promote homeostasis.

FCC = Face-centered cubic crystalline
A structure in which the simplest repeating unit consists of fourteen equivalent lattice points, eight of which are at the corners of a cube and another six in the centers of the faces of the cube. Found in cubic closest-packed structures.

Femto = 10^{-15} second

Feynman (ϕ) = Nanometer (*nm*)
Nanoscale unit of length for the first time proposed in the present book in honor of Richard P. Feynman, the original advocate of nanoscience and nanotechnology. {One Feynman (ϕ) $\equiv$ 1 Nanometer (*nm*)= 10 Angstroms (Å)= 10^{-3} Micron (μ) = 10^{-9} Meter (*m*)}.

Fibrinogen
A protein present in blood plasma involved in coagulation. Fibrinogen converts to fibrin and reacts with other molecules to produce blood clots.

F1 =Fluorophore group
An atomic group with one excited molecule that emits photons and is fluorescent.

FISH =Fluorescent-In-Situ Hybridization

FLAPW = Full-Potential Linearized Augmented Plane Wave.
An all-electron method which within DFT is universally applicable to all atoms of the periodic table of elements and to systems with compact as well as open structures.

Fluorescent Probe
Fluorescence Microscopy is a widely used technique in biology to observe and identify molecular composition of structures through the use

of fluorescently-labelled probes of high chemical specificity such as antibodies.

FRACTAL
A mathematical construct that defines non-Euclidean geometric dimension of certain random structures named "FRACTAL objects.

Fullerene   (See also **Buckminsterfullerene**)
Cage-like structures of carbon atoms, There are fullerenes containing 60, 70, 76, ... to 500 carbon atoms.

Gaussian
A computer package of *ab initio* programs.

Glycoside   •
Derived from glucose and are common in plants, but rare in animals.

Glu   = Glutamine

Hamiltonian
Has two distinct but closely related meanings. In classical mechanics, it is a function, which describes the state of a mechanical system in terms of position and momentum variables, which is the basis of Hamiltonian mechanics. In quantum mechanics, the Hamiltonian refers to the observable corresponding to the total energy of a system.

Hartree-Fock
A self-consistent iterative procedure to calculate the so-called "best possible" single determinant solution to the time-independent Schrödinger equation. As a consequence to this, whilst it calculates the exchange energy exactly, it does not calculate the effect of electron correlation at all. It is only applicable after the Born-Oppenheimer approximation has been made.

HCP   Hexagonal Close Pack crystalline structure.
HCP structure is a special case of a hexagonal structure with alternating layers shifted so its atoms are aligned to the gaps of the preceding layer, and with $c/a = sqrt(8/3) = 1.633....$ With this ratio, the atomic separation within a layer (on the x-y plane) is the same as the separation between layers.

HCV   = Hepatitis C Virus

Heisenberg Uncertainty Principle
The quantum mechanical principle of Werner Heisenberg "The more precisely the POSITION is determined, the less precisely the MOMENTUM is known".

Hessian
The square matrix of second partial derivatives of a function (assumed to be twice differentiable).

HIS   = Histidine

HIV   = Human Immunodeficiency Virus

Host-guest chemistry
Construction of molecular receptors by self-assembly so that such receptors could, to some extent, gain molecular recognition capability.

IgG   = Immunoglobulin G

IMIC   = Immobilized Metal Ion Complexation

Immuno-PCR   = I-PCR
A tool for sensitive protein quantification. A combination of the enormous amplification power of polymerase chain reaction (PCR) with antibody-based immunoassays. Allows the detection of proteins at a level of a few hundred molecules.

Inclusion Complex (See **Inclusion Compound**)

Inclusion Compound   =Inclusion Complex
The mechanical trapping of small molecules within spaces between other molecules. A compound in which one component (the host) forms a cavity or, in the case of a crystal, a crystal lattice containing spaces in the shape of long tunnels or channels in which molecular entities of a second chemical species (the guest) are located. There is no covalent bonding between guest and host, the attraction being generally due to van der Waals forces. If the spaces in the host lattice are enclosed on all sides so that the guest species is "trapped" as in a cage, such compounds are known as "clathrates" or "cage" compounds".

Intermolecular Forces
Forces due to attraction and repulsion between molecules.

Intermolecular Potential
Potential energy due to attraction and repulsion between molecules.

Kohn-Sham Equations
Lagrange-multiplier equations for density functional theory.

Lagrangian
Defined to be the difference between the kinetic energy and the potential energy of an object. Lagrangian mechanics is a re-formulation of classical mechanics introduced by J. L. Lagrange in 1788. In Lagrangian mechanics, the trajectory of an object is derived by finding the path which minimizes the action, a quantity which is the integral of the Lagrangian over time.

LAPW = Linearized Augmented Plane Wave
A method for solving the equations of density functional theory (DFT). DFT) has proven to be an accurate and reliable basis for the understanding and prediction of a wide range of materials properties from the first principles of quantum mechanics (ab initio), without any experimental input.

LCAO = Linear Combination of Atomic Orbitals
The most common formalism for building molecular spin orbitals is the LCAO method.

LDL = Low Density Lipoprotein
A molecule which is a combination of lipid (fat) and protein. Lipoproteins are the form in which lipids are transported in the blood. LDL transports cholesterol from the liver to the tissues of the body.

Lectin
Naturally produced proteins or glycoproteins that can agglutinates (united as if by glue) cells. It can also bind with carbohydrates or sugars to form stable complexes.

Ligand
A radical, an ion, a molecule, or a molecular group that binds to another chemical entity to form a larger complex.

Ligation
The act of binding or of applying a ligature; to ligate; See DNA ligation.

Linear Response Theory
Originally introduced by Kubo [R. Kubo, *J. Phys. Soc. Japan*, **12**, 570, (1957)]. It is valid close to equilibrium. Forms the theoretical basis for evaluation of transport properties by molecular dynamics simulation. It is consistent with Lars Onsager's linear theory of irreversible thermodynamics.

Lipoprotein
An important class of serum proteins in which a spherical hydrophobic core of triglycerides or cholesterol esters surrounded by an amphipathic monolayer of phospholipids, cholesterol and apolipoproteins.

Liposome
Synthetic, relatively uniform bilayer lipid membrane-bound vesicles formed by emulsification of cell membranes in dilute salt solutions. Liposomes are being developed as an approach for drug delivery in which relatively toxic drugs are "wrapped" inside a liposome and tagged with an organ-specific antibody.

LISA = Lithography Induced Self-Assembly

LSDA = Local Spin Density Approximation

Magnetostriction
The phenomenon wherein ferromagnetic materials experience an elastic strain when subjected to an external magnetic field.

MBB = Molecular Building Block.

MEMS = MicroelectroMechanical Systems
A generic term to describe micron scale electrical/mechanical devices.

Metropolis algorithm
In this algorithm, one can start from an arbitrary point in [a,b), say $x_0=(a+b)/2$, and add a random number to it $x_1=x_0+d(2r-1)$ where r is the random number in [0,1) and d is the magnitude of the step. If $P(x_1)/P(x_0)>1$, then the move is accepted; else the ratio is compared to a random number in [0,1).

Markov Process
A stochastic process whose future probabilities are determined by its most recent values and not on how it arrived in the present state. The

random walk towards equilibrium from an arbitrary initial state, with no memory from previous steps.

Micelle
The nanostructure formed by amphipathic molecules in solution that places the polar group toward the solution and the hydrophobic group toward the interior. Colloid particle composed of many aggregated small molecules having a layered structure.

Microelectrophoresis
Microscopic observation of the movement of a single suspended particle through a fluid under the action of an electromotive force applied to electrodes in contact with the fluid.

Mitochondria
A small intracellular organelle which is responsible for energy production and cellular respiration.

MNT = Molecular Nanotechnology
Control of the structure of matter based on intermolecular interactions control to perform molecular manufacturing and generate molecular machinery.

Moiety
A portion of a molecular structure having some property of interest.

Molecular Dynamics (MD)
This consist of the study of intramolecular conformations and molecular motions, using computer simulation based on the equations of motion and energy of a finite number of particles and their resulting statistical averaging.

Molecular Recognition
Used in combinatorial chemistry. A chemical term referring to processes in which molecules adhere in a highly specific way, forming a larger structure.

MOLPRO
A computer package of *ab initio* programs.

Monoclonal Antibody
An antibody that consists of a single type of immunoglobulin since it is produced artificially from a single cell clone.

Monte Carlo Simulation (MC)
A stochastic computation / simulation technique based on random number generation of the variable to estimate the function statistically.

mRNA
Messenger Ribonucleic Acid. Controls the synthesis of new proteins.

Nanobiotechnology
Nanotechnology applications in biological systems. Development of technology to mimic living bio systems.

Nanocatalysis
Present day production of catalysts is by tedious and expensive trial-and-error in laboratory in large-scale reactors. The catalytic action occurs on surface of highly dispersed ceramic or metallic nanostructures. Nanotechnology facilities may bring about a more scientific way of designing new catalysts named nanocatalysis with precision and predictable outcome.

Nanocrystal
Orderly crystalline aggregates of 10s-1000s of atoms or molecules with a diameter of about 10 nm.

Nanolithography
Writing in the nanoscale.

NanoManipulator
Application of special virtual reality (VR) goggles and a force feedback probe (as an interface) to a scanning probe microscope like AFM or STM.

Nanomanufacturing   = molecular manufacturing

Nanomaterial
Refers to nanoparticles, nanocrystals, nanocomposites, etc. The bottom-up approach to material design.

Nanoscale
One billionth of meter scale.

Nanostructure
Geometrical structures in nanoscale

Nanosystem
Controlled volume or controlled mass systems defined in nanoscale.

NLS = Nuclear Localization Signal

nm = nanometer = ϕ (as **Feynman**)

NMDA = N-Methyl-D-Aspartate

Nucleocytoplasmic Transport
The directed movement of molecules between the nucleus of a living cell and the cytoplasm. (this is considered to be a biological process ontology).

Nucleic acid
A large molecule composed of nucleotide subunits. DNA and RNA are major types of Nucleic acids.

Nucleoprotein
Any of several substances found in the nuclei of all living cells; consists of a protein bound to a nucleic acid.

Nucleotides
The subunits of a DNA molecule. Monomers of which DNA and RNA are built. They consist of phosphates, a nitrogenous base and a carbon chain.

Octet
A set of eight valence electrons in an atom or ion, forming a stable configuration.

Oligomer molecule
A molecule of intermediate relative molecular mass, the structure of which essentially comprises a small plurality of units derived, actually or conceptually, from molecules of lower relative molecular mass.

Oligonucleotide
A molecule usually composed of 25 or fewer nucleotides; used as a DNA synthesis primer.

Oligopeptide
A molecule consisting of a string of four to nine amino acid units.

Oncogene
A gene that that can cause a cell to develop into a tumor cell. Many oncogenes are involved, directly or indirectly, in controlling the rate of cell growth.

Optimization

A design (or mathematical) procedure in which a proper (or an extremum) conditions of a system (or variables of a function) is sought.

Organelle
Any part of a cell that has a unique structural, functional or anatomical role. On a smaller scale, organelles are similar to the organs in your body - they are, in effect, the organs of a cell.

Ontology
An explicit formal specification of how to represent the objects, concepts, and other entities that are assumed to exist in some area of interest and the relationships that hold among them.

Oxime
Any of a group of compounds containing a CNOH group, formed by treating aldehydes or ketones with hydroxylamine.

PCR = Polymerase Chain Reaction
The process that makes a huge number of copies of a gene.

PEG = Poly Ethylene Glycol

PEO = Poly Ethylene Oxide

Peptide
A short compound formed by linking two or more amino acids. Proteins are made of multiple peptides.

Peptide Bond
A chemical bond formed between two molecules when the carboxyl group of one molecule reacts with the amino group of the other molecule, releasing a molecule of water. This is a dehydration synthesis reaction, and usually occurs between amino acids.

Phage
A virus that is parasitic in bacteria; it uses the bacterium's machinery and energy to produce more phage until the bacterium is destroyed and phage is released to invade surrounding bacteria.

Phage Display
The process by which the phage is made to 'display' human antibody proteins on its surface.

Phagocytosis
Endocytosis of particulate material, such as microorganisms or cell fragments.

Phonon
A quantum of sound, usually observed in a crystal. It is to sound exactly what a photon is to light. In the lattice vibrations of a crystal, the phonon is a quantum of thermal energy. It is given by $h.f$, where h is the Planck constant and f the vibrational frequency.

Pico $= 10^{-12}$ second

Plasmid
A structure in cells consisting of DNA that can exist and replicate independently of the chromosomes.

PNA =Peptide Nucleic Acid

Poisson's Equation

A second-order partial differential equation, $\nabla^2 \varphi = f$, or $\nabla^2 V = -\dfrac{\rho}{\epsilon_0}$

$\Delta\varphi = f$, i.e., it sets the Laplacian equal to f. Finding φ for some given f is an important practical problem, since this is the usual way to find the electric potential for a given charge distribution.

Positional Assembly
Constructing materials one atom or molecule at a time using nanomanipulators.

PPV = poly(p-phenylenevinylene)
A semiconducting conjugated polymer (in short known as PPV polymer)

Primeval Soup
The atmosphere of the earth while it was being formed some 5000-600 million years ago, consisting, first, mostly of methane, ammonia and carbon dioxide. Later on water, then amino acids, then purines and pyridines, then RNA and DNA, then simple bacteria-like creatures and algae and then more complex creatures began to appear at the beginning of the Cambrian period, 570 million years ago.

pRNA = Pyranosyl Ribonucleic Acid

Prodrug
An inactive form of a drug that exerts its effects after metabolic processes within the body converts it to a usable or active form.

Prophylaxis
Prevention of disease, or the preventive treatment of a recurrent disorder.

Protein Folding
The process that proteins acquire their functional preordained three-dimensional structure after they emerge, as linear polymers of amino acids, from the ribosome.

Prophylaxis
Treatment intended to prevent the occurrence of disease.

Proteases
Any of various enzymes, including the endopeptidases and exopeptidases, that catalyze the hydrolytic breakdown of proteins into peptides or amino acids.

Proteome
All the proteins expressed by a genome.

Proteomics
It is about identification of proteins in the body and the determination of their role in physiological and pathophysiological functions.

Quantum Mechanics
The science to describe a system of particles in terms of a wave function defined over the configuration of particles having distinct locations. Schrödinger equation has a the leading role in quantum mechanics analogous to Newton's second law in classical mechanics.

Quantum Dots
Nanometer-sized solid state structures made of semiconductor or metal crystals capable of confining a single, or a few, electrons. The electrons possess discrete energy levels just as they would in an atom.

Quantum Well
Consists of a thin layer of a narrower-gap semiconductor between thicker layers of a wider-gap material.

Quantum Wire
Unlike the single-dimension "quantum dot," it is in two dimensions - It has "length," and allows the electrons to propagate in a "particle-like" fashion. Usually built on a semiconductor base, and (among other things) used to produce very intense laser beams of up to multi-giga-Hertz per second.

Rayleigh-Ritz variational principle
A technique for computing eigenfunctions and eigenvalues.

Regioselectivity
A regioselective reaction is one in which one direction of bond making or breaking occurs preferentially over all other possible directions. Reactions are termed completely (100%) regioselective if the discrimination is complete.

Relaxation Time
The time required for a system to recover a specified condition after a disturbance. In stochastic processes it is a measure of the rate at which a disequilibrium distribution decays toward an equilibrium distribution.

Restriction Digestion
The process of cutting DNA molecules into smaller pieces with special enzymes called Restriction Endonucleases (also called Restriction Enzymes or RE's). RE's recognize specific sequences in the DNA molecule.

Ribosome
A naturally occurring molecular machine that manufactures proteins according to instructions derived from the cell's genes.

RME
Receptor Mediated Endocytosis

Rhodopsin
Consists of the protein opsin linked to 11-cis retinal a prosthetic group. Retinal is the light absorbing pigment molecule and is a derivative of vitamin A. Opsin is a member of the 7TM (transmembrane) receptor family.

RNA
Ribonucleic acid - a single-stranded nucleic acid made up of nucleotides. RNA is involved in the transcription of genetic information; the information encoded in DNA is translated into messenger RNA (mRNA.).

SAM
Self-Assembled Monolayer.

Scaffold   (see **Template**)

Scanning Tunneling Microscope
An instrument able to image conducting surfaces to atomic accuracy; has been used to get images and pin molecules to a surface.

S-Layer = Surface Layer

Schrödinger equation

$$i\hbar\frac{\partial\Psi}{\partial t} = -\frac{\hbar^2}{2m}\frac{\partial^2\Psi}{\partial x^2} + V(x)\Psi(x,t) \equiv \tilde{H}\,\Psi(x,t),$$

Describes the time-dependence of quantum mechanical systems. Has a role in quantum mechanics analogous to Newton's second law in classical mechanics $i = \sqrt{-1}$, $\hbar = h/(2\pi)$, h is Planck's constant, Ψ is wavefunction, t is time, m is mass, x is position, $\hat{H}$ is Hamiltonian, V is potential energy. It also provides a quantitative description of the rate of change of the "state vector" which encodes the probabilities for the outcomes of all possible encodes the probabilities for the outcomes of all possible. State vector represents the set of all possible states of a system described by a complex Hilbert space. Any instantaneous state of a system is described by a unit vector in that space. In quantum mechanics, this state vector measurements applied to the system. As the state of a system generally changes over time, the state vector is a function of time.

Self-assembly = Brownian assembly
In atomic and molecular level it refers to random motion of atoms and molecules and the affinity of their binding sites for one another.

Self-replication =exponential replication (see **Von Neumann Machine**)

Small System
A system defined in nanoscale (nanosystem)

Sonication
The process of disrupting biologic materials by use of sound wave energy.

SPM = Scanning probe microscope

Steepest descent method = gradient descent method
An algorithm for finding the nearest local minimum of a function, which presupposes that, the gradient of the function can be computed.

Stewart Platforms
A positional device as described in detail in Positional (Robotic) Assembly part of this book.

Sui generis
(Latin). Having a distinct character of its own; unlike anything else.

STM
Scanning tunneling microscope.

STM
Scanning tunneling microscope. This is an instrument able to image conducting surfaces to atomic accuracy; has been used to pin molecules to a surface. This is a more powerful device than AFM.

STV = Streptavidin

Supramolecule
A complex of molecules formed as a result of intermolecular (non-covalent) interactions and not covalent bonds. DNA is an example of supramolecules.

Superposition
A quantum mechanical phenomenon in which an object exists in more than one state simultaneously.

Superposition Approximation
In statistical mechanics it is referred to Kirkwood's proposed approximation for the triplet correlation function and its relation to pair correlation functions.

TB = Tight-Binding
A quantum mechanical tool originally proposed by J. C. Slater and G. F. Koster [Phys. Rev. 94, 1498 (1954)]. In TB the wavefunction is expanded in terms of a set of localized states $m=1,...,M$ in each atomic layer j. A variation of the TB scheme is the "Wannier Orbital Model".

Thermal Noise
In atomic and molecular level it refers to vibration and motion of atoms and molecules at temperatures above absolute zero.

Tropism
The movement response of an organism to an external stimulus.

Trypsin Hydrolysates
The result of catalytic hydrolysis of a protein by trypsin (an enzyme of pancreatic origin) as the catalyst.

Tumor necrosis factor
A type of biological response modifier (a substance that can improve the body's natural response to disease).

Template
Core portion of a molecule common to all members of a combinatorial library in combinatorial chemistry. A molecule, such as DNA, that serves as a pattern for the synthesis of a macromolecule, as of RNA.

THF = Tetrahydrofuran

T_g = Glass Transition Temperature
The temperature at which characteristics of a polydispersed material change from that of a glass (amorphous solid) to that of rubber. Below T_g molecules have very little mobility.

Therapeutic
Treatment and care to combat disease or alleviate pain or injury.

TNF = Tumor Necrosis Factor

Translation (in genetics)
The process whereby genetic information coded in messenger RNA directs the formation of a specific protein at a ribosome in the cytoplasm.

Transcoytosis
Process of transport of substance across an epithelium (the cells or membrane covering the outside of organs) by uptake into and release from coated vesicles.

Tropism
The turning or bending movement of an organism or a part toward or away from an external stimulus.

Variational principle
Based on variational calculus. An approach to solving a class of optimization problems that seek a functional to make some integral function an extreme. In statistical mechanics it is used to find an approximate solution to the partition function. In quantum mechanics it provides the starting point for almost all methods which aim to find an approximate solution to Schrödinger's equation.

Vasopressin
A medication that causes a contraction of blood vessels.

Vesicle

A closed membrane shell, derived from membranes either by a physiological process (budding) or mechanically by sonication. Vesicles of dimensions in excess of 50 ϕ [*nm*] are believed to be important in intracellular transport processes.

Von Neumann Machine

A machine that is able to build a working replica of itself using materials in its environment. See von Neuman's paper about self-reproducing machines at [von Neumann, J., 1966, The Theory of Self-reproducing Automata, A. Burks, ed., Univ. of Illinois Press, Urbana, IL].

VON Neumann Probe

A von Neumann Machine able to move over interstellar or interplanetary distances and to utilize local materials to build new copies of itself. Such probes could be used to set up new colonies, perform megascale engineering or explore the universe.

Voronoi Polyhedron

The region of space around an atom, such that all points of this region are closer to this atom than to any other atom of the system.

Wavefunction

A wavefunction, Ψ, is a result of solving the Schrödinger's equation in quantum mechanics. It is a scalar function that is used to describe the properties of a wave. Its boundary conditions include the fact that it must be continuous across boundaries and also have continuous derivatives. Wavefunctions must also be normalized such that $\int\limits_{-\infty}^{+\infty} |\psi|^2 dx = 1$, where $|\psi|^2$ stands for the distribution function.

Wet Nanotechnology

For organic nanosystems that exist primarily in a water environment and living systems. Nanoscale structures of interest here include enzymes, genetic material, membranes, and other cellular components.

Zeolite

A natural or synthetic hydrated aluminosilicate with an open three-dimensional crystal structure.

Zinc-finger
A finger-shaped fold in a protein, which permits it to interact with RNA and DNA. It is created by binding of specific amino acids in the protein to a zinc atom. Zinc-finger proteins regulate the expression of genes as well as nucleic acid recognition, reverse transcription and virus assembly.

Index